AF577404

HEAVY METAL POLLUTION RESEARCH
Recent Advances

HEAVY METAL POLLUTION RESEARCH
Recent Advances

Editor
Professor (Dr.) Arvind Kumar
Environmental Science Research Unit
Post Graduate Department of Zoology,
S.K.M. University, Dumka – 814 101 (Jharkhand)

2006
DAYA PUBLISHING HOUSE
Delhi - 110 035

ISBN 81-7035-385-5

Published by : **Daya Publishing House**
1123/74, Deva Ram Park
Tri Nagar, Delhi – 110 035
Phone: 27383999
Fax: (011) 23244987
e-mail : dayabooks@vsnl.com
website : www.dayabooks.com

Showroom : 4762-63/23, Ansari Road, Darya Ganj,
New Delhi – 110 002
Phone: 23245578, 23244987

Laser Typesetting : **Classic Computer Services**
Delhi – 110 035

Printed at : **Chawla Offset Printers**
Delhi – 110 052

PRINTED IN INDIA

Preface

We have all been warned about the man-made toxic substances that are now documented to exist in the air we breathe, the water we drink and the food we eat but how many of us have taken our thought process one step further by asking ourselves the critical question. Papers recently published in prestigious journals finally confirm that low levels of heavy metals (even at levels that were once considered "safe") are in fact, most definitely dangerous. This new research documents the effects of these toxic metals as they accumulate in the body over time, and clearly indicates that Ca-EDTA chelation therapy provides benefits by reducing the body's burden of these toxic heavy metals, resulting in improved physiological functioning and much better health.

Heavy metals such as lead poisoning has long been recognized as a health hazard. Lead has been historically used in a number of industrial processes, including manufacturing batteries, paints, and adding it to gasoline. Acute (high exposure) lead poisoning causes symptoms of abdominal pain or "lead colic"cognitive deficits, peripheral neuropathy, arthralgias, decreased libido, and anemia. Lead is practically everywhere in today's environment. It enters our bodies from many sources including defective glazes (pottery), drinking water, contaminated soil, airborne particulate, leaded gasoline, paint and several other sources. Symptoms of lead poisoning are stomach pains, constipation, diarrhea, aggressiveness, anxiousness, hyperactivity, muscle pain, weakness, weight loss, learning disabilities, convulsions and eventual death with chronic lead poisoning. Lead poisoning victims usually become anemic. These symptoms usually persist for about 2 weeks from time of exposure, and then settle into the organs, bones & even hair. Unfortunately, we still do not know the long-term effects of lead exposure. Likewise long term exposure to even low levels of any heavy metals may result in development of many disorders and different diseases. Keeping this adverse impact of heavy metal pollution, I have ventured to compile some recent research articles of eminent toxicologists of India in order to aware the generations to come.

My special thanks and appreciation go to the scientists whose contributions have enriched this volume. I wish to express my sincere gratitude to Dr. P. C. Hembram, Hon'ble Vice Chancellor, S. K. M. University, Dumka who has been a source of constant inspiration. I am especially thankful to Professor (Dr.) B. N. Jha, Hon'ble Pro Vice-Chancellor, S. K. M. University, Dumka for his encouragement. I owe my special thanks to Professor M. C. Dash, Hon'ble Vice Chancellor of Sambalpur University, Professor N. C. Datta of Calcutta University, Professor S. K. Konar of Kalyani University, Professor P. S. Murthy of Bangalore University, Professor P. Natarajan of Trivandrum University, Professor S. P. Roy of Bhagalpur University, Professor A. K. Quereshi of Bhopal University, Professor Tanmay Bhattacharya of Midnapore University, Professor P. C. Mishra of Sambalpur University, Professor B. D. Joshi of Hardwar University, Professor G. C. Pandey of Faizabad University, Professor K. C. Sharma of Ajmer University, Professor M. Raziuddin of Hazaribag University, Professor U. S. Bagde of Mumbai University, Dr. M. P. Sinha of Ranchi University, Professor Gurdeep Singh of I. S. M., Dhanbad, Dr. P. K. Goel of Karad, Dr. M. C. Varma and Shri T. Poddar of Bhagalpur University. I also acknowledge the incentives provided by one of my research scholars, Dr. Chandan Bohra, Head of Zoology, B. S. K. College, Barharwa. for helping me actively in bringing out this book.

I also express my deep sense of gratitude to my parents whose blessings have always prompted me to pursue academic activities deeply. I am also thankful to my sweet wife, Professor Kumari Bimla and my two lovely sons, Kumar Pallav Shivshankaran and Kumar Prasun Ramakrishnan whose natural smiles extended to me relief all through this tiresome endeavour.

Last but not the least, I am also thankful to Mr. Anil Mittal, Proprietor, Daya Publishing House, Delhi for taking keen interest in bringing out of this book. Finally, I will always remain a debtor to all my well-wishers for their blessings, without which this book would not have come into existence.

Dumka **Professor Arvind Kumar**

Contents

List of Contributors

A. Subramanian

P.G. and Research Department of Zoology, Khadir Mohideen College, Adirampattinam - 614701, Tamil Nadu, India

A.A. Sivakumar

P.G. and Research Department of Zoology, Kongunadu Arts and Science College, Coimbatore - 641 029, Tamil Nadu, India

A.K. Misra

Professor, Chemistry Department, Gauhati University, Guwahati - 7810 14, Assam

A.K. Panigrahi

Laboratory of Environmental Toxicology, Department of Botany, Berhampur University, Berhampur - 760 007, Orissa

A.N. Subramaniyan

Center of Marine Environmental Studies, Ehime University, Japan

Abhijit Mitra

Department of Marine Science, University of Calcutta, 35, B.C. Road, Kolkata - 700 019

Anumita Das

Department of Marine Science, University of Calcutta, 35, B.C. Road, Kolkata - 700 019, W.B.

Arvind Kumar

Environmental Science Research Unit, Post Graduate Department of Zoology, S.K.M., University, Dumka – 814 101, India

B. Nabi

Department of Biological Sciences, Pondicherry University, Pondicherry - 605 014, India

Bharat Bhusan Patnaik

Research Scientist, Environmental Sciences and Biotechnology Research Unit, P.G. and Research Department of Zoology, Loyola College (Autonomous), Chennai - 600 034

C. Bohra

Department of Zoology, B.S.K. College, Barharwa (Sahibganj), Jharkhand

C. Sheela Sasikumar

P.G. Department of Biochemistry, D.G Vaishnav College, Chennai - 106, India

D. Anusuya

Postgraduate Department of Zoology, Voorhees College, Vellore - 632 001, India

D.J. Prakash

Postgraduate Department of Zoology, Voorhees College, Vellore - 632 001, India

D.P. Bhattacharyya

Department of Theoretical Physics, Indian Association for the Cultivation of Science, 2A and 2B Raja S.C. Mullick Road, Jadavpur, Kolkata - 700 032, India, E-mail: tpdpb@mahendra.iacs.res.in

Debarati Mukherjee

Department of Theoretical Physics, Indian Association for the Cultivation of Science, Jadavpur, Kolkata - 700 032, W.B.

H. Ali

Post Graduate Department of Chemistry, S.K.M. University, Dumka – 814 101 (Jharkhand)

H.S. Sharma

Indira Gandhi Centre for H.E.E.P.S., University of Rajasthan, Jaipur - 302 004

I. Christy

Postgraduate Department of Zoology, Voorhees College, Vellore - 632 001, India

J. Hongray Howrelia

Research Scholar, Environmental Sciences and Biotechnology Research Unit, P.G. and Research Department of Zoology, Loyola College (Autonomous), Chennai - 600 034

Jyotsna Lal

Chemistry Department, Christ Church Post Graduate College, Kanpur - 208 001, U.P. India, Email: jyotsna_lal@yahoo.com

K. Muthukumaravel

P.G. and Research Department of Zoology, Khadir Mohideen College, Adirampattinam - 614701, Tamil Nadu, India

K. Sasikumar

Department of Botany, Annamalai University, Annamalai Nagar - 608 002 Tamil Nadu, India

K. Srikumar

Department of Biological Sciences, Pondicherry University, Pondicherry - 605 014, India

K.M. Kulkarni

Environmental Physiology Laboratory, P.G. Department of Zoology, Government Vidarbha Mahavidyalaya, Amravati - 444 604

Kakoli Banerjee

WWF-India Secretariat, Tiger and Wildlife Programme, Canning Field Office, South 24 Parganas - 743 329, W.B.

Kanchan Sachan

Chemistry Department, Christ Church Post Graduate College, Kanpur - 208 001, U.P. India

Lingaraj Patro

Laboratory of Environmental Toxicology, Department of Botany, Berhampur University, Berhampur - 760 007, Orissa

M. Aruchami

P.G. and Research Department of Zoology, Kongunadu Arts and Science College, Coimbatore - 641 029, Tamil Nadu, India

M. Choudhary

Department of Zoology, M.L.S.M. College, Darbhanga- 846 004, Bihar

M. Selvanayagam

Professor, Environmental Sciences and Biotechnology Research Unit, P.G. and Research Department of Zoology, Loyola College (Autonomous), Chennai - 600 034

M. Srinivasan

CAS in Marine Biology, Annamalai University, Parangipettai - 608 502 Tamil Nadu, India

M.I.S. Saggoo

Department of Botany, Punjabi University, Patiala - 147002, Punjab, India

M.K. Paul

Senior Lecturer, Chemistry Department, Lumding College, Lumding - 782 447, Assam

M.M. Jha

Department of Zoology, M.L.S.M. College, Darbhanga- 846 004, Bihar

Meenakshi Mishra

Laboratory of Environmental Toxicology, Department of Botany, Berhampur University, Berhampur - 760 007, Orissa

N. Mayilvakanan

P.G. and Research Department of Zoology, Khadir Mohideen College, Adirampattinam - 614701, Tamil Nadu, India

N. Verma

Department of Biotechnology, Punjabi University, Patiala - 147002, Punjab, India

P. Martin Deva Prasath

Reader, P.G. Department of Chemistry, TBML College, Porayar - 609 307, Tamil Nadu, South India

P. Sharma

Department of Botany, Punjabi University, Patiala - 147002, Punjab, India

Paulomi Maiti

Aquaculture Research Unit, Department of Zoology, University of Calcutta, 35, Ballygunge Circular Road, Kolkata - 700 019

R. Rajaram

CAS in Marine Biology, Annamalai University, Parangipettai - 608 502 Tamil Nadu, India, E-mail: drrajaram69@ahoo.com

R. Thangam

P.G. and Research Department of Zoology, Kongunadu Arts and Science College, Coimbatore - 641 029, Tamil Nadu, India

R.B. Dhake

B/H Maharana Pratap High-School, A/P and Tal.-Bhusawal - 425 201, District Jalgaon (MS)

Rajib Chakraborty

Department of Marine Science, University of Calcutta, 35, B.C. Road, Kolkata - 700 019, W.B.

Richa Marwari

Indira Gandhi Centre for H.E.E.P.S., University of Rajasthan, Jaipur - 302 004

S. Anitha

P.G. Department of Biochemistry, D.G Vaishnav College, Chennai - 106, India

S.S. Patil

Department of Environmental Science, Dr. Babasaheb Ambedkar Marathwada University, Aurangabad - 431 004

S.T. Ingle

School of Environmental and Earth Science, North Maharashtra University, Jalgaon - 425 001, M.S., India

S.V. Deshmukh

Department of Biology, Rural Institute (Agriculture), Amravati - 444 603

S.V.S. Amanulla Hameed

P.G. and Research Department of Zoology, Khadir Mohideen College, Adirampattinam - 614701, Tamil Nadu, India

Samir Banerjee

Aquaculture Research Unit, Department of Zoology, University of Calcutta, 35, Ballygunge Circular Road, Kolkata - 700 019, E-mail: samirban@vsnl.net

Sutapa Das

Department of Environmental Science, University of Calcutta, 35, B.C. Road, Kolkata - 700 019

T. Mandal

Department of Marine Science, Calcutta University, 35 B.C. Road, Kolkata - 100 019, India

T.I. Khan

Indira Gandhi Centre for H.E.E.P.S., University of Rajasthan, Jaipur - 302 004

V. Mahesh

P.G. Department of Biochemistry, D.G Vaishnav College, Chennai - 106, India

Chapter 1

Heavy Metal Pollution: Toxic Effects and Control Strategies

Arvind Kumar, C. Bohra and H. Ali

Introduction

Metals are ubiquitous in nature and with increasing industrialization, the potential for metallic poisoning is increasing day by day. A metal in trace amount, less than 0.01 per cent is essential and in the absence of that metal, an organism fails to grow or complete its life cycle. However, the same trace metals may prove to be toxic when the concentration level exceeds these required for correct nutritional response by factors between 40 and 200 folds (Venugopal and Luckey, 1975). Thus, an undersupply of trace metals leads to a deficiency, sufficient supply results in optimum conditions and over supply results in the toxic effects and lethality at the end.

Modern research has led to a broader understanding that metal ions have a biological significance, in contrast to classical concept that inorganic chemistry is restricted to non-living chemical system, whereas the living world falls within the realm of organic and bio-chemistry. Experimental studies have proved that the role of heavy metal ions in living systems follows the pattern of natural availability (Vehrenkamp, 1973: Wood, 1974). There is no life, that can survive without the participation of metal ions. Some of the major ions such as Na, K, Mg, Ca are essential to sustain life. In the same way some of the heavy metals are considered essential both for plants and animal nutrition and they serve some useful biological functions in the body. Thus, Cu, Co, Mn, Mo, Se and Zn are essential to both plants and animals. However, some trace metals such as Cd, Hg, Pb and metalloids like As, Sb and Se are considered to be toxic, although some essential physiological roles have been inferred for As, Cd and Pb recently (Dara, 1993).

Toxicity by heavy metals is induced by delivery of the metals to the cell and the specific action of the metal determines the ultimate severity of the toxic action. High natural concentrations of metals in food and water could have led to the first exposure. Metals leached from eating utensils or metallic cookware increase the risk of exposure. Intentional use of compounds containing toxic metals as pesticides or as therapeutic agents increase the opportunity for hazardous exposure. The advent of industrial era led to more widespread occurrence of occupational diseases related to exposure to a wide variety of toxic metals. In recent years a justifiable concern has arisen with regard to pollution of environment by toxic metals.

The heavy metals are important component of pollutants which not only cause phytotoxicity but also enter into the food chain causing hazardous impacts on human health and animals. The phytotoxic impacts of heavy metal pollution are very commonly observed on crops such as chillies, rice, tomatoes etc. Water, food and soil are essential for life but when contaminated they transmit bacteria, virus and parasites that cause some of the world's most menacing diseases such as diarrhoea, cholera, typhoid, intestinal worms, hepatitis, tetanus and cancer.

Exposure to different metals may occur in common circumstances particularly in an industrial setting. Major sources of chronic and low level exposure are smelting and refining process due to fume inhalation. Plating, alloy formation and various other forms of metal cooking may lead to acute and high level exposure to metals and their inorganic compounds. Toxicity by organic compounds can result during their manufacture, handling or secondary use. Metals and their salts are also used therapeutically such as mercurials as diuretics, lithium carbonate in the treatment of psychiatric disorders and bismuth for gastrointestinal distress. However, the metals that are essential nutrients can exert toxic action in concentrations above physiological limits. The metals having the greatest potential for causing diseases are those which accumulate in the body.

Although excessive concentrations of metals may occur in water, air or soil as a result of natural deposits, technologic use of these non-biodegradable materials can lead to their accumulation in the environment. Metals released to the environment may be bio-accumulated and thus enter the food chain. Mercury compounds released with industrial water may be converted by microbial systems in water beds to the highly toxic methyl mercury which is taken up by fish leading to fatal diseases in consumers (Hammond and Beliles, 1980). In the industrial situation, inhalation is the most important route of exposure. Some metals like Ni, Be and Ar induce skin changes as part of their toxicity. Topical exposure to certain occupational metals may result in irritation of the skin and eyes or sensitization reactions and provide a route of absorption resulting systemic toxicity.

Today, heavy metal pollution is a great problem to the environment. The groundwater contamination particularly by the industrial effluents (heavy metals) and their persistence in food chains has been of major concern as it is posing a serious threat to aquatic cultures including fisheries. Today, the heavy metals are termed as "devils in disguise". These are bio-accumulative and relatively stable as well as toxic/carcinogenic and therefore, require close monitoring.

Water pollution by heavy metals exert alarming effects on aquatic organisms (Timmerman *et al.*, 1992). These heavy metals produce changes in the chemistry of water, in turn, change the life of aquatic organisms (Thirumathal *et al.*, 2002). In India, cadmium, lead, mercury, cobalt etc. have been very common heavy metals and their residues have been reported in water resources (Gupta *et al.*, 2002). Apart from acute toxicity the heavy metals in water has proved dangerous and harmful because of their bio-accumulation (Panda and Sahu, 2002) and their impact on tissue degeneration. These heavy metals are also found to exert a definite toxic effect on phytoplankton which caused increased respiration and decreased photosynthesis and thus, decreased primary productivity (Bohra and Kumar, 2003). Singh *et al.* (2003) also studied heavy metal pollution in some natural waterbodies of Jharkhand.

Factors Influencing Toxicity

In elemental form, metals rarely react with biological systems. But soluble salts of metals dissociate readily in the aqueous environment of biological membranes, thereby facilitating their transport as metal ions. Conversely insoluble salts are relatively poorly absorbed. The absorption of soluble metallic salts are modified by certain factors. Some soluble salts are rendered insoluble in presence of an anion. For example, dietary phosphate reduces absorption of lead by forming insoluble lead phosphate. Foods also bind with many metals reducing their absorption. Metals like methyl mercury and tetraethyl lead occur in nature as alkyl compounds which remain intact in the body even after absorption and distribution. Most toxic metals strongly bind to body tissues and accumulate in the body. Lead, radium and strontium have strong affinity for bones and cadmium and mercury deposit in the kidney.

Chelation of Metals

Metals form stable complexes with certain agents. This forms the basis of enhancing their excretion by administering a readily excreted coupling agent. The use of citrate was suggested as a means of promoting lead excretion (Kety, 1942). Later dimercaprol was developed as an antidote for arsenic poisoning. Dithiol compounds, especially those with SH groups on adjacent carbon atoms, were found to give the best protection for arsenic poisoning. Chelating agents are generally non-specific in their affinity for metals. In addition to mobilizing and excreting toxic heavy metals, they also enhance the excretion of essential metals like calcium and zinc. So, certain metals in the body compete with toxic metals for the binding sites in chelating agents. Another widely used chelating agent is ethylene diamine tetra acetic acid (EDTA) which is used as an antidote for lead poisoning.

Mode of Action

There is no generalized mechanism by which toxic metals produce their effects. Different metals affect different target organs and systems. They have different affinities for binding at various sites in the body. Due to this multiplicity of effects and diversity of actions, the concept of critical organ and critical dose is used for metals (Nordberg, 1976).

Lead (Pb)

Lead is the most intensively studied metal from toxicologic point of view. Lead in the environment exists almost entirely in inorganic form, but small amount of organic lead results from the use of leaded gasoline (Trin, 1979). It is highly toxic either as the inorganic (various salts and oxides of lead) or organic (alkyl lead) compound. It is found in soil, vegetation, animals, food, water and air in the vicinity of highway. However, its level decreases exponentially with the distance from the road (Sharma, 1995). The natural dispersion of lead in the environment is somewhat restricted due to its insolubility. However, overall human exposure to leads is primarily from food. The use of lead in various industries including pesticides is the other sources, which contribute lead in the environment. Lead in gasoline accounting about 20 per cent of lead used by mankind is responsible for about 98 per cent of pollution problem. Lead levels in sewage sludge may vary between 2000 to 8000 ppm and its use as a fertilizer may give rise to subsequent contamination of agricultural soils.

Source

The highest level of exposure occurs in workers from occupational environment. In the general population, the major hazard is for young children who swallow objects contaminated with lead containing paints. Lead poisoning results from ingestion of lead based paints which contain triplumbic tetra-oxide (red lead), basic lead carbonate (white lead), lead sulphate or lead chromate. Other sources include used oil, grease, linoleum, leaded petrol solid lead solder, roofing material, asphalt, discarded storage batteries and lead arsenate in white lotion. Atmospheric pollution results from fumes or wastes from lead smelting or plumbing works. Grass near busy highway may contain toxic amounts of lead from auto exhausts. Lead is one of the most common causes of poisoning in dog and cattle. Special areas of exposure occur in printing accumulator batteries, paint and plastic industries.

The atmospheric lead concentration in high density urban area may be 3 $\mu g/m^3$ (Krishnan, 1995). However, WHO (1973, 1977) suggest that the lead level in ambient air around some industrialized areas is as high as 6 $\mu g/m^3$. Deposited lead from the air contaminate the soil. In highly contaminated soil the lead level is of the order of 2 g/kg. However, WHO (1977) has recorded lead level of over 10 g/kg. The lead content in natural lake and river water of the world has been estimated to be 1–10 $\mu g/l$. It is estimated by number of agencies that a typical daily dietary intake of lead ranges from

100–500 μg/day (Voege, 1971; Meranger and Smith, 1972; WHO, 1976; NRC, 1977; USPEA, 1989). The world wide average dietary intake of lead for an adult is about 200 μg/day. For different individuals and population groups the exposure of lead from water, food, air etc. can vary significantly.

Absorption

The major routes of lead absorption is Gastro-intestinal tract (GIT) and some absorption of lead occurs through intact or abraded skin when applied in high concentration (Laug and Kunze, 1948; Rastogi and Clausen, 1976). Organic lead compounds are absorbed through skin of workers causing toxicity. The average dermal lethal dose of tetra ethyl lead is 700 mg/kg Pb in rabbits which is about six times only the oral lethal dose (Kehoe, 1927). Intestinal absorption is low due to low solubility. Lead mostly enters the body through mouth. The absorption of lead from the gastro intestinal tract is greatly influenced by concurrent dietary levels of calcium, iron, fats and proteins (Barltrop and Khoo, 1975). Exposure to the dust of the metal ($LbCl_2$, $PbBr_2$ or PbO_2) leads to high absorption, as much as 90 per cent if the particles are less than 0.1 μm diameter. Absorption is considerably greater in infants and children than in adults, which is thought to be due to differences in the nature of diet (Kostial *et al.*, 1971; Forbes and Reina, 1972). The overall absorption of lead taken as inadvertent contaminants of foods and beverages by adults is approximately 8 per cent (Kehoe, 1961; Rabinowitz, 1974) in contrast to 40 per cent in children (Ziegler *et al.*, 1978). Lead tetraethyl is readily absorbed through the lungs and skin. Krishnan (1995) suggested that about 50 per cent of inhaled airborne lead of particle size (0.9 μm) is readily absorbed through the lungs and if the particle size is still small (0.1 μm) than about 90 per cent of inhaled lead may be absorbed. On an average about 10 per cent of inorganic lead is ingested from food and water is absorbed. The amount of lead ingested by a person through food depends on:

1. The total amount of food eaten,
2. The growth history of food,
3. The opportunity of foods to absorb lead from water, soil and tropical deposition, and
4. Personal dietary habit.

Distribution

Absorbed lead enters the blood and is distributed to soft tissues and bones. The half life of lead in blood is 1 to 2 months and steady state is thus achieved in about t six months. However, the concentration of lead in bone appears to increase (Gross *et al.*, 1975) and half life in bone is estimated to be 20 to 30 years. The residual fraction of lead becomes progressively buried deeply in the bone matrix with time and becomes inaccessible to the blood circulation. Lead accumulates in bone throughout the life (Barry, 1975 and Gross *et al.*, 1975). About 90 per cent of the total body burden of lead with long term exposure is deposited in bones kidney and liver also contain high lead concentration. Lead also enters the brain both in the inorganic form and more readily the organic form due to its solubility in fats (Gupta *et al.*, 2002).

Toxicity

Lead is understood today to cause adverse effects at the levels of exposure that produce no clinically detectable symptoms and that only a few years ago were thought to be safe (ATSDR, 1993). This recognition of sub clinical toxicity first arose from studies in young children showing that chronic asymptomatic exposure to lead could cause irreversible injury to the nervous system (Needleman, 1979; Bellinger, 1987; McMichael, 1988; Dietrich, 1991 and Wasserman, 1992). At low doses, lead has a generalized stimulatory effect with increased DNA and protein synthesis cell respiration and production of RBC. An important point is that lead toxicity has a cumulative effect.

The toxic effect of lead are produced by a variety of different mechanisms:

1. It inhibits the sulphahydryl groups of vital enzymes of cellular metabolism,
2. It interferes with Cu, Fe and Zn as prosthetic groups of enzymes in mitochondria thus affecting cellular respiration, oxidative phosphorylation and ATP synthesis,
3. It damages blood brain barrier thus causing the cytotoxic effects of solutes in the brain,
4. It causes rupture of lysosome and release of acid phosphatase,

5. It interferes with the flux of solutes such as calcium into and out of the brain,
6. It inhibits Na^+, K^+ and ATPase enzymes in the RBC and other cells,
7. It inhibits both adrenergic and cholinergic synaptic neurotransmission (Kostial and Vouk, 1957; Manalis and Cooper, 1973; Cooper and Steinberg, 1977), and
8. It inhibits haem synthesis thus causing haemoglobin deficiency.

Toxic Effects

The four major targets are the central nervous system, peripheral nerves, kidneys and the haematopoietic system. Acute toxicity is uncommon except for exposure to high concentration of Pb dust.

Effects on Central Nervous System

Lead encephalopathy is a fatal condition due to chronic or sub-chronic exposure to high dose of inorganic lead resulting from increased solubility of lead in lipids causing demyelination in cerebrum, spinal cord and peripheral nerves which lead to hyper activity and insomnia. The major features are dullness, restlessness, irritability, headache, muscle tremors, ataxia and loss of memory. These signs may progress to convulsions coma and death. A high incidence of residual damage is seen including epilepsy, hydrocephalus and idiocy. In fatal cases cerebral damage occurs commonly along with demyelination and axonal damage in neurons. At lower exposure level, lead causes behavioral effects particularly in children. Organic compounds produce psychic disturbances like hallucinations, delusions and excitement which progress to delerium in fatal cases. Acute CNS symptoms include paraesthesias pain and muscle weakness. The CNS syndrome is usually more common among children (Klaasen, 1991).

Effect on Peripheral Nervous System

Lead plasy frequently occurs by occupational exposure in workers. The major clinical manifestations are weakness of extensor muscles and sensory disturbances liker hypertension and analgesia. The lesions of this peripheral neuropathy are characterized by segmental demyelination and axonal degeneration. Functionally nerve conduction velocity is decreased.

Thus, lead is toxic to the central and peripheral nervous system as well as kidney dysfunction occurs due to the impairment of energy metabolism, leading to the expression of Fanconi Syndrome (Krishnan, 1995).

Effect on Kidney

Two types of renal effects are produced by lead. In the first type, there are manifestations of damage to the proximal tubules. Tubular re-absorption of glucose, amino acids and phosphate is hindered. The second type of effects occur with prolonged high lead exposure. It is a progressive condition characterized by interstitial fibrosis, sclerosis of vessels and glomerular atrophy.

Effect on Haemato Poiesis and Heme Synthesis

Anaemia is one of the early manifestations of lead poisoning which results from reduced life span of circulating RBC and inhibition of haemoglobin synthesis. Effects on heme synthesis are more important than the effects on globin synthesis.

Other Effects

Chronic exposure to lead may also cause weight loss, constipation and loss of teeth. It may cause naucea abdominal pain and vomiting. If large amount of lead is absorbed rapidly, a shock syndrome may be developed as a result of massive gastrointestinal loss of fluid. It also produces effects like chromosomal aberrations and abnormal sperm morphology in man. The gums often slow a blue line due to deposition of colloidal $Pb_3 (PO_4)_2$. Growth may be impaired and longevity is reduced. Infections may occur because of suppressed immune response. Lead nephrosis is caused due to mitochondrial damage in renal tubules when the blood level of lead exceed 150 μg/ml.

Treatment

The major specific therapeutic objective is removal of lead from the body using chelating agent. Sodium or magnesium sulphate should be given orally to remove any lead remaining in the intestine. D-penicillamine can be given orally to chelate the lead in the gut. In adults the most widely accepted procedure is intravenous infusion of the calcium salt of disodium ethylene diamine tetra acetate (Ca EDTA). 1–2 g daily for 4–5 consecutive days to mobilize lead from the tissues and enhance its urinary excretion. Lead is mobilized mainly from bones (Hammand, 1971). The treatment in children includes a course of EDTA therapy, either alone or in combination with dimercaprol. Combined therapy is more effective than therapy with either drug alone (Chisolm, 1970). Thiamine may be give simultaneously as a supportive therapy.

Mercury (Hg)

Mercury is a unique pollutant because of its apparent indestructibility and ubiquitousness. It attracted worldwide attention as a potential pollutant because of a tragic episode which occurred in the Japanese villages surrounding the Minamata Bay, The victims were struck with a mysterious nervous illness commonly called "Minamata disease" which was later shown to be caused by consuming fish and shell fish that has lethal accumulation of methyl mercury (Takeuchi, 1968). It resulted in 121 cases of poisoning and 41 death. A search ultimately traced the source to be the waste discharge containing high amounts of methyl mercury and other organic mercurials from industries using mercury as catalyst in the manufacture of vinyl chloride and acetaldehyde. Mercury appears in industrial discharges in five principal forms such as divalent mercury (Hg^{2+}), metallic mercury (Hg), phenyl mercury (C_6H_5Hg), alkoxyalkyl mercury (Clt_3–O–Clt_2–CH_2–Hg^+) and methyl mercury (Clt_3 Hg^+) of which methyl mercury is the most toxic (Mitra, 1986).

Concentration of mercury in air, soil and water have increased because of greater use of fossil fuels and its industrial and agricultural use.

Source

Broadly, there are two major sources of mercury from which it enters into the environment. First, mercury present in the earth's crust gets released into the environment as a result of physical or chemical erosion and degassing. This release of mercury is not due to man's action and hence it as a natural phenomenon. The second source of mercury in the environment is anthropogenic in nature involving direct or indirect result of man's action. The important user of mercury is the electrical apparatus industry which manufacture mercury batteries, alkaline mercury cells, mercury pool rectifiers, power tubes, fluorescent and photocopying lamps etc. Laboratory instruments such as mercury switched, relays, diffusion pumps, valves, pump seals, barometers, thermometers and vibration dampers largely use mercury. Paint industries consume a large quantity of mercury, mostly organomercurials to give the paint anti-microbial and anti-fungal properties. Dental preparations.and other clinical sources utilize substantial quantity of mercury. Certain chemical industries use mercury as a catalyst, for example, during conversion of acetylene into acetaldehyde, vinyl chloride and vinyl acetate. Mercury is also used in paper and pulp industry as a slimicide. Though, the use of mercury as insecticide and fungicide has decreased greatly due to severe restrictions or ban over its use, a large part of mercury is still used in the manufacture of pesticides and fungicides. Mercury is also used in pharmaceuticals and cosmetics. Mercurus chloride or calomel is used as antiseptic, diuretic and cathartic. Major incidents of human poisoning occur from consumption of mercury treated seed grains. Human poisoning have also occurred by ingestion of meat of pigs which had been fed grain treated with organomercurial fungicide.

Mechanism of Action

Mercurials even in low concentrations inactivate sulphahydryl enzymes thus interfering with cellular metabolism and function. Mercury also combines with phosphoryl carboxyl, amide and amine groups. Both inorganic and alkyl mercury disrupt the integrity of blood brain barrier affecting brain metabolism. Degenerative changes are wide spread in the cell bodies, and nerve fibres. Sensory neurons are more severely affected than moter neurons. Degeneration of both the axoplasm and the myelin sheath is common. Inhibition of enzymes of glycolytic pathway and protein synthesis may be the cause of degenerative changes. Various neuro-physiologic parameters are also altered by mercury. It blocks synaptic and neuromuscular transmission. High concentration of mercury vapour causes direct irritation of the lungs and inorganic mercury causes inflammation of the gastrointestinal mucosa.

Absorption

Elemental mercury vapours when inhaled is completely absorbed by lung. Its per cent deposition and retention in man quite high (upto 80 per cent) because of its monoatomic nature and lipid solubility. However, in gastro intestinal tract, it is very poorly absorbed (less than 0.01 per cent) because it occurs as large globular particles in GIT. After absorption, in RBC, it is rapidly oxidized to divalent form (Hg^{2+}) by an enzyme, catalase. The disposition of inhaled elemental mercury is of special interest because of the importance of intoxication by this route. A significant amount of vapour enters the brain before being oxidized and produces CNS toxicity (Magos, 1967). Upon prolonged inhalation of mercury, its urinary excretion exceeds faecal excretion. The urinary excretion is roughly proportional to the level of air exposure.

Inorganic mercury salts in food are absorbed up to 10 per cent after ingestion. Some absorption of inorganic mercury occurs from skin. Organic mercurials are more completely absorbed from the intestine. The absorption of methyl mercury, even mixed with food, is about 95 per cent in adults. Poisoning from alkyl mercurials has been known to occur from dermal application of methyl mercury ointments. Toxic manifestations of inorganic mercury are renal whereas those of methyl mercury are neurologic (Sahu *et al.*, 2002).

Toxic Effects

Mercury produces toxic effects on many organs and systems. Depending upon the chemical form, the major target organs affected are the central nervous system and the kidney. The effects of mercury vapour exposure are neuropsychiatric whereas, those of methyl mercury are of sensorimoter nature. Acute exposure to mercury vapour may produce symptoms such as weakness, chills, metallic taste, naucea, vomiting, diarrhoea, dyspnoea, cough and pneumonitis. Chronic exposure to mercury vapour mainly produces neurological effects. There may be goitre, tachycardia, tremor, depression, irritability, insomnia, severe salivation, gingivitis and renal dysfunction.

Kidney is the primary target organ only of inorganic mercury. High doses cause severe acute poisoning characterized by anuria progressing to polyuria, grey mucosa of mouth pharynx and intestine, intense pain, vomiting, hypovolemic shock, strong metallic taste stomatitis and renal tubular necrosis. In several cases, disturbances in tubular function may persist for several months after poisoning (Valek, 1965). Chronic exposure to inorganic mercury is predominated by glomerular injury, erythema of extremities, chest and lace with photophobia, anorexia and tachycardia as a result of hypersensitivity reaction to mercury.

Symptoms of exposure to organic mercurials are mainly neurological including visual disturbances, ataxia, hearing loss.

vestibular dysfunction, loss of sensor of smell and taste, mental deterioration, muscle tremor, incoordination, paralysis and in severe cases death.

Treatment

Immediate termination to exposure, prompt attention to fluid and electrolyte balance, gastric lavage to remove mercury from the alimentary tract are some of the measures for the treatment of mercury poisoning. Recommended treatment includes dimercaprol (5 mg/kg i.m. followed by 2.5 mg/kg i.m.) every 12 hours for 10 days. Penicillamine 250 mg orally every 6 hours, may be used alone or following dimercaprol treatment. According to Bakir *et al.* (1976), the clinical efficacy of penicillamines in the treatment of methyl mercury poisoning is not satisfactory because the neurologic effects of these compounds result from irreversible damage to neurons.

Cadmium (Cd)

Cadmium appears to be highly toxic because of the absence of homeostatic control for this metal in human body. It is a metal of current toxicological concern. It occurs in nature in association with Zn and Pb. Extraction of processing of Zn and Pb often lead to environmental contamination with Cd. Local and other fossil fuels contain cadmium and their combustion releases the element into the environment. It has wide application in electroplating, galvanization, plastics, paints and nickel cadmium batteries. Workers in smelters and other metal processing plants are exposed to high concentration of Cd in the air cadmium and agricultural soil is mainly derived from fertilizer, fungicides and sewage sludge applied in the crop field. The root of corn oat, soyabean, tomato, alfa alfa accumulates highest level of Cd when grown in Cd contaminated soil. The aerial part of carrot, potato, lettuce and tomato accumulate highest level of Cd. Thus, water, food and smoking are the major sources of cadmium for general population.

Human beings are exposed to cadmium either through inhalation or by the ingestion of contaminated foods and drinks.

Absorption

Cadmium as readily absorbed through ingestion or through the lungs. About 1-2 per cent of ingested and about 11 per cent of inhaled Cd is retained in the body. Elementary absorption is affected by a number of factors such as age, Cd, Fe, Zn and protein deficiency (CES, 1972, 1979 and USEPA, 1980) and the chemical form of the cadmium ingested. Dietary factors, such as Fe, Ca and protein deficiency, may increase the gastro intestinal absorption rate (Flangan *et al.*, 1978). In iron deficient woman, up to 20 per cent of ingested cadmium was found to be absorbed. The active absorption of calcium from small intestine requires calcium binding protein (Ca B P). Increased Ca BP activity induced by calcium deficiency enhances cadmium absorption (Washko and Cousins, 1977). Young animals absorb cadmium to a much greater extent than old ones. During chronic exposure, more than 90 per cent of circulating cadmium in the blood cells partially bound to metallothionein and partially to haemoglobin. Cadmium has a strong preferential affinity for liver and kidney. Pulmonary absorption depends on the size and solubility of the particles containing Cd and also affected by the depth and rate of breathing. Approximately 50 per cent of 0.1 µm particles will be retained by the lung in contrast to 20 per cent for particles as large as 2 µm (CES, 1979). Absorbed cadmium will enter the blood and become concentrated in certain parts of the human body (CES, 1978 and Underwood, 1971). Both the liver and kidney act as the organs for the accumulation of cadmium and about 50 per cent of any accumulated cadmium is found in these organs (CES, 1978). Cadmium is highly cumulative and progressively accumulates in human body upto the age of 50 years.

Toxic Effects

The toxic effects of long term exposure of cadmium may somewhat vary with the route of exposure. The kidney is affected following either pulmonary or gastrointestinal exposure (Klaassea, 1991). It is associated with induction of hypertension, cirrhosis of liver etc. (Higgins and Burns, 1975). It is also linked with pulmonary emphysema and lung cancer. In the case of oral intake naucea, vomiting salication, diarrhoea and abdominal cramps are common symptoms of cadmium poisoning (Klaassen, 1991). Death may occur due to shock and dehydration. Kidney is affected following any not of exposure but effects on lung are seen only after exposure by inhalation. Symptoms include glycosuria, hypercalciuria, increased uric acid excretion and proteinuria due to renal injury (Kajantzis *et al.*, 1963 and Adams *et al.*, 1969), dysponea, emphysema, pulmonary fibrosis osteomalacia and hyperlycaemia.

Treatment

Respiratory support and steroid therapy is helpful. Chelation therapy is $CaNa_2$ EDTA at the rate of 75 mg/kg/day in three to six divided doses for five days. Large doses of vit. D given over a period of months are effective in relieving painful symptoms and reduce the incidence of spontaneous fractures.

Copper (Cu)

It is widely distributed in nature. It occurs in several oxides, carbonate and sulphide ores as well as native copper. Copper is an essential trace element required by oxidative enzymes such as catalase, peroxidase and cytochrome oxidase and for haemoglobin synthesis. The common sources of copper to animals and man include spray of copper containing compounds an plants, absorption by plants from soil, contaminated plants grown in the vicinity of mines or smelters, drinking water from copper pipes, use of copper sulphate as fungicides etc. The role of this trace element in human health is a complete enigma, while some studies have shown unequivocally that a deficiency of copper can lead to high serum cholesterol and an increased risk of cardiovascular, disease (Reiser *et al.*, 1987). The recommended daily intake of copper ranges from 2–3 mg/l for human adults. The ingestion of 15–75 mg of copper causes gastrointestinal disturbance.

Absorption

Copper is absorbed from the intestine. It is initially bound to serum albumin and later more firmly to alpha ceruloplasmin. Most of the copper from blood is accumulated in liver and bone marrow. It binds tightly to the SH groups of cytoplasmic proteins chelatin and metallothionein. The liver excretes copper in bile but there is reabsorption in the intestine.

Toxicity

Copper toxicity or deficiency in animals not only merely dependent upon copper intake but also depends upon dietary levels of Zn, Fe and Ca (Fostner and Wiltmann, 1981). Continuous ingestion of copper from food or water reduces chromic copper poisoning in man and causes a kind of disease called "Wilsons disease" which is characterized by excessive copper concentration in the tissues arising from metabolic defects involving absorbed copper and not from ingestion of excessive amount of copper. Clinical signs includes haemoglobinuria, jaundice and bloody nasal discharge. There are signs of respiratory in sufficiency and shock. Enlargement of liver, kidney and spleen, distended gall bladder containing thick greenish bile, excess pericardial fluid lung edema, dark musculature and spongy degeneration of brain are some common symptoms. Generally copper toxicity could be aggravated by low dietary Mo or Zn (Krishnan, 1995). Other work suggests that excess copper causes neurological complaints, hypertension, liver and kidney dysfunction cancer and accelerated aging (Pfeiffer and Mailloux, 1987). Acute copper poisoning causes naucea, vomiting, salivation, diarrhoea, violent abdominal pain, convulsions, hypotension, paralysis, collapse and death. Lesions include marked gastroientestine, centrilobular hepatic neurosis (Chuttani *et al.*, 1965) with congestion of liver, spleen and kidney.

Treatment

It is done by administration of ammonium molybdate and sodium sulphate orally.

Selenium (Se)

Selenium is obtained as a byproduct of copper refining. It is used in electronics industry for rectifiers, photocells and solar batteries, steel, paint, varnish, glass and ceramic manufacturing units and as a vulcanizing agent for rubber. Selenium dioxide is used as an oxidizing agent and as an oxidant in lubricating and other oils. Selenium poisoning or Selenosis can occur from improper use of selenium containing shampoos used in the treatment of dermatitis or too much of selenium intake.

Source

The main source of selenium are the plants that grow in selenium rich soils. It accumulates in certain plants in sufficient quantities to produce toxicity in livestocks. Cereals, vegetables and other food crops grown on such soils contain dangerous levels of selenium. A concentration of 5 ppm in food or 0.5 ppm in milk or water has been estimated to be dangerous for human beings. Seleniferous regions are generally characterized by low rainfall. Although, selenium deficient diets cause a number of diseases such as liver necrosis in rats, multiple organ necrosis in mice, pancreatic fibrosis and alopecia in chicks, stiff lamb disease and necrotic liver degeneration and cardiac myopathy in young pigs, but its excess intake causes toxicity. Selenium toxicity can also occur in man by consuming eggs of affected birds.

Absorption

It is readily absorbed from the intestine and is distributed throughout the body. Its highest concentration is found in kidney, liver and spleen while the lowest is found in brain and muscle. After prolonged administration, its large amount is found in hair and nails. Selenium is transmitted across the placenta and hence congenital abnormalities occur in the new born. Toxicity occurs when intake exceeds the excretory capacity (McConnel and Portman, 1952 and Schroeder *et al.*, 1970).

Toxicity

The reaction of selenium with thiol groups may be the primary source of selenium toxicity (Diplock, 1976). Acute toxicity produces CNS affects like nervousness drowsiness and convulsions. Other clinical signs include a rapid and weak pulse, dysponea, colic pain, polyuria, cyanosis and death due to respiratory failure. Sub-acute poisoning occurs due to ingestion of seleniferous plants. It is characterized by impairment of vision, weakness of limbs and respiratory failure (Moxan and Rhian, 1943). There is weight loss, staggening gait, salivation, lacrimation, severe abdominal pain, inability to swallow and finally complete paralysis, collapse and death. Liver is atrophied, necrotic and spleen is enlarged. Chronic poisoning is characterized by dullness, loss of vitality, emaciation and depraved appetite. There may be increasing lameness due to erosion of the articular surface of long bones and atrophy of hooves. Signs of intoxication may also include decayed teeth, skin eruptions, gastrointestinal distress and partial loss of hair and nails. Symptoms of chromic inhalation exposure may include gastrointestinal disorders, nervousness, garlic breath, liver and spleen damage, anaemia, mucosal irritation and lumbar brain. Selenium has produced loss of fertility and congenital defects and is considered embryotoxic and teratogenic (Moxan and Rhian, 1943; Schroeder *et al.*, 1970 and Robertson, 1970). Lesions include atrophy of heart, cirrhosis, gastroenteritis and nephritis.

Treatment

Sub-acute poisoning by selenium may be treated by injection of strychnine if the paralytic stage has not reached. Feeding small amounts of arsenic protects against toxic doses of selenium by binding it is the liver in a non-toxic form. High protein diet and food rich in Vit. E are used to prevent its toxicity (Sellers *et al.*, 1950).

Arsenic (As)

Arsenic is found in soil, water and air as a common environmental toxicant. Arsenic is found in high concentration in some sources of drinking water. It is also found in coal and released in air during combustion. Application of arsenic containing insecticides, weed killers, food preservatives and phosphate fertilizers has increased its environmental dispersion. Arsenic is a natural contaminant of food. Sea foods, park liver and salt may be exceptionally high in arsenic.

Absorption

Arsenical compounds may be absorbed after ingestion or by inhalation. Rate of absorption from the digestive tract depends on solubility. Sodium arsenite is readily soluble, rapidly absorbed and highly toxic. Arsenic trioxide is least soluble and slowly absorbed. Arsenic accumulates in the liver from where it is slowly released and distributed in other tissues. It is then stored for a long time in bones and skin and permanently in keratinized tissues like hair and nails. Toxicity of arsenic is due to its trivalent compound.

Toxicity

Symptoms include intense abdominal pain, staggering gait, collapse and paralysis. Arsenic poisoning is usually acute with major action on gastrointestinal tract and cardio-vascular system. Symptoms include profuse watery diarrhoea, severe colic pain, dehydration, weak pulse, marked skeletal muscle cramps, shock, convulsions, coma and death occurs within 24 hrs. Chromic poisoning is characterized by muscle weakness, fatique, skin pigmentation, hyper-keratosis, edema, itching, jaundice, anaemia renal dysfunction, peripheral neuritis and paralysis of extremities. There is red inflammation of intestinal mucosa, edema of submucosa, perforation of gastrointestinal wall and foul smelling.

Treatment

Chelation therapy is done initially with dimercaprol at the rate of 3 mg/kg intramuscularly every 4 hours until abdominal symptoms subside. After this penicillamine is given orally upto 1 g/day for

4 days. In severe cases of nephropathy renal dialysis may be performed (Srivastava and Dumka, 2001).

Conclusion

The industry either should be closed or strict instructions should be given not to discharge the effluent into the river and the estuary. The solid waste should not be dumped outside. The effluent and waste should be treated properly and toxicants should be recovered and recycled. The industry must conduct weekly healthy check up camps and the patients should be supplied with adequate medicines and adequate food by the industry owner to the same children from unnatural death due to malnutrition and environmental contaminants. The industry must follow the rules and guidelines setup by the Pollution Control Board. The Pollution Control Board instead of sitting as a silent spectator must rise to the occasion and implement environment Act strictly to the true sense. Then only a safe environment can be presented to the future generation. For better living "A Clean Environment" is needed.

References

Adams, R.G., Harrison, J.F. and Scott, P. (1969). The development of cadmium-induced proteinuria impaired renal function and osteomalacia in alkaline battery workers. *Quart. J. Med.*, 38: 425–433.

ATSDR (1993). Toxicological profile for lead. TP 92–12 Atlanta GA: Agency for toxic substances and Disease registry.

Bakir, F., Al-Khalidi, A., Clarkson, T.S. and Greenwood, R. (1976). Clinical observations on treatment of alkymercury poisoning in hospital patients. In: Conference on Intoxication Due to Alkyl mercury-treated Seed. *Bull. WHO (Suppl.)*, 53: 87–92.

Barltrop, D. and Khoo, H.E. (1975). The influence of nutrition factors on lead absorption. *Postgrad. Med. J.*, 51: 795–800.

Barry, P.S.I. (1975). A comparison of concentrations of lead in human tissues. *Br. J. Ind. Med.*, 32: 119–139.

Bellinger, D., Leviton, A., Waternawx, C. and Needleman, H.L. (1987). Longitudinal analysis of prenatal and postnatal exposure and early cognitive development. *N. Engl. J. Med.*, 315: 1037–1043.

Bohra, C. and Kumar, A. (2003). Impact of mercury cadmium and lead on phytoplankton productivity in a lentic aquatic environment. In: *Environment Pollution and Management*, (Eds.) Kumar, A., Bohra, C. and Singh, L.K. APH Publ. Corp., Delhi, p. 419–422.

CES (Commission of European Communities) (1972). *International Symposium on Environmental Health Aspects of Lead*. Amasterdam, Proceedings, pp. 1168.

CES (Commission of European Communities) (1978). *Criteria Dose/ Effects Relationship for Cadmium*. Oxford Pergamon Press.

CES (Commission of European Communities) (1979). *Trace Metal: Exposure and Health Effects*. Oxford Pergamon Press.

Chisolm, J.J. (1970). Treatment of acute lead intoxication choice of chelating agents and supportive therapeutic measures. *Clin. Toxicol.*, 3: 527–540.

Chuttani, H.K., Gupti, P.S. and Gultati, S. (1965). Acute copper sulphate poisoning. *Am. J. Med.*, 39: 849–854.

Cooper, G.P. and Steinberg, D. (1977). Effects of cadmium and lead on adrenergic neuromuscular transmission in the rabbit. *Am. J. Physiol.*, 232: 128–131.

Dara, S.S. (1993). *A Textbook of Environmental Chemistry and Pollution Control*. S. Chand and Co. Ltd., Ram Nagar, New Delhi, pp. 210.

Dietrich, K.N., Succop, P.A., Berger, O., Hamnond, P. and Bornschain, R.L. (1991). Lead exposure and cognitive development of urban preschool children: The Cincinnati lead study cohort at age 4 years. *Neurotoxicol Teratol.*, 13: 203–211.

Diplock, A.T. (1976). Metabolic aspects of selenium action and toxicity. *Crit. Rev. Toxicol.*, 4: 271–329.

Flanagan, P.R. (1978). Increased dietary Cd absorption in mice and human subjects with Fe deficiency. *Gastroenterology*, 74: 841–851.

Forbes, G.B. and Reina, J.C. (1972). Effect of age on gastrointestinal absorption (Fe, Sr and Pb) in the rat. *J. Nutr.*, 102: 647–652.

Forstner, U. and Wittmann, G.T.W. (1981). *Metal Pollution in the Aquatic Environment*. Springer-Verlag, Berlin. Heidelberg, New York.

Gross, S.B., Pfitzer, E.A., Yeager, D.W. and Kehoe, R.A. (1975). Lead in human tissues. *Toxitol. Appl. Pharmacol.*, 32: 638–651.

Gupta, D.C., Pefers, E. and Yadava, R.N. (2002). Concentration of heavy toxic metals in waters of the area around Mandideep industrial complex and its impact on human health. In: *Ecology of Polluted Waters*, Vol. I, (Ed.) Kumar, A. APH Publ. Corp., Delhi, p. 515–536.

Hammond, P.B. (1971). The effects of chelating agents on the tissue distribution and excretion of lead. *Toxicol. Appl. Pharmacol.*, 18: 296–310.

Halllmond, P.B. and Beliles, R.P. (1980). Metals. In: *Casarett and Doull's Toxicology: The Basic Science of Poisons*, (Eds.) Doull, J., Klassen, C.D. and Amdur, M.O. MacMillan Publ. Co. Inc., New York, p. 409–467.

Higgins, I.J. and Burns, R.G. (1975). *The Chemistry and Microbiology of Pollution*. Academic Press, New York. p. 189–210.

Kajantzis, G., Flynn, F.V., Spowage, J.S. and Trott, D.G. (1963). Renal tubular malfunction and pulmonary emphysema in cadmium pigment workers. *Quart. J. Med.* 32: 165–192.

Kehoe, R.A. (1927). On the toxicity of tetra ethyl lead and inorganic lead salts. *J. Lab. Clin. Med.*, 12: 554–560.

Kehoe, R.A. (1961). The metabolism of lead in man in health and disease. The Harben Lectures, 1960. *J.R. Inst. Pub. Health. Hyg.*, 24: 101–120.

Kety, S.S. (1942). The lead citrate complex ion and its role in the physiology and therapy of lead poisoning. *J. Biol. Chem.*, 142: 181–192.

Klaassen, C.D. (1991). Heavy metals and heavy metal antagonists: The Pharmacological basis of therapeutics. Vol. II, 8th ed. Pergamon Press Inc., Singapore, 66: 1592–1614.

Kostial, K. and Vouk, V.B. (1957). Lead ions and synaptic transmission in the superior conical ganglion of the cat. *Br. J. Pharmacol. Chemother.*, 12: 219–222.

Kostial, K., Simonovic, I. and Pisonic, M. (1971). Lead absorption from the intestinal in newborn rats. *Nature*, 233: 564–574.

Krishnan, K. (1995). *Fundamentals of Environmental Pollution*. S. Chand and Co. Pub. Ltd., pp. 166.

Laug, E.P. and Kunze, F.M. (1948). The penetration of lead through the skin. *J. Ind. Hyg. Toxicol.*, 30: 256–259.

Magos, L. (1967). Mercury-blood interaction and mercury uptake by the brain after vapour exposure. *Rev.*, 1: 323–337.

Manalis, R.S. and Cooper, G.P. (1973). Presynaptic and postsynaptic effects of lead at the frog neuromuscular function. *Nature*, 243: 354–355.

McConnell, K.P. and Portman, O.W. (1952). Excretion of dimethyl selenide by the rat. *J. Biol. Chem.*, 195: 277–282.

McMichael, A.J., Baghurswt, P.A., Wigg, N.R., Vimpani, G.V., Robertson, E.F. and Roberts, R.J. (1988). Port, pine cohort study. Environmental exposure to lead and children's abilities of four years. *N. Engl. J. Med.*, 319: 468–475.

Meranger, J.C. and Smith, D.C. (1972). The heavy metal content of a typical Canadian diet. *Canadian J. Pub. Hlth.*, 63: 53–65.

Mitra, S. (1986). *Mercury in the Ecosystem*. Trans. Tech. Publ. Ltd., Switzerland.

Moxan, A.L. and Rhian, M. (1943). Selenium poisoning. *Physiol. Rev.*, 203: 305–337.

Needleman, H.L., Gunnoe, C. and Leviton, A. (1979). Deficts in psychological and classroom performance of children with elevated dentine lead levels. *N. Engl. J. Med.*, 300: 689–695.

Nordberg, G.F. (1976). *Effects and Dose-Response Relationships of Toxic Metals*. Elsevier Sci. Pub. Co., New York.

NRC (National Research Council) (1977). *Drinking Water and Health*. National Acad. Sci., Washington, D.C.

Panda, D.S. and Sahu, R.K. (2002). Heavy metal pollution in a tropical lagoon, Chilka lake. *Ind. J. Env. and Ecoplan.*, 6: 39–43.

Pfeiffer, C.C. and Mailloui, R.C. (1987). Excess copper as a factor in human disease. *J. Phy. Med. NMR.*, 16: 175–195.

Robinowitz, M. (1974). Lead contamination of the biosphere by human activity: A stable isotope study. *Ph.D. Thesis*, California Univ.

Rastogi, S.C. and Clausen, J. (1976). Absorption of lead through the skin. *Toxicol.*, 6: 371–376.

Reiser, S., Powell, A., Yang, C.Y. and Canary, J.J. (1987). Effect of copper intake on blood-cholesterol and in lipoproteins distribution in men. *Nutr. Rep. Int.*, 36: 641–647.

Roberston, D.S.E. (1970). Selenium a possible teratogen. *Lancet.*, 1: 518–519.

Sahu, A., Panigrahi, M.K., Sahu, S.K. and Panigrahi, A.K. (2002). Aquatic mercury pollution by a chlor-alkali industry and its impact on the ecology of human health of Ganjam area: A case study. In: *Ecology of Polluted Waters*, Vol. I, (Ed.) Kumar, A. APH Publ. Corp., Delhi, p. 465–496.

Schroeder, H.A., Forst, D.V. and Balassa, J.J. (1970). Essential trace metal in man: Selenium. *J. Chromic. Dis.*, 23: 227–243.

Sellers, E.A., You, R.W. and Lucas, C.C. (1950). Lipotropic agents in liver damage produced by selenium or carbon tetrachloride. *Proc. Soc. Exp. Biol. Med.*, 75: 118–121.

Sharma, D.C. (1995). Essentiality and toxicity of certain trace elements: A strategy. In: *Environmental Strategies*, (Ed.) Lodha, R.M. Himanshu Publ., Udaipur.

Singh, L.K., Roy, P.N. and Kumar, A. (2003). Heavy metal pollution in some natural water bodies of Jharkhand. In: *Environment Pollution and Management*, (Eds.) Kumar, A., Bohra, C. and Singh, L.K. APH Publ. Corp, New Delhi, p. 423–428.

Srivastava, A.K. and Dumka, V.K. (2001). Environmental pollution due to heavy metals and their control strategies. In: *Current Topics in Environmental Sciences*, (Eds.) Tripathi, G. and Pandey, G.C. ABD Publ., Jaipur, p. 234–257.

Takeuchi, T. (1968). *Pathology of Minamata Diseases*, (Ed.) Kutsuma, M. Japan, pp. 141.

Thirumathal, K., Sivakumar, A.A., Chandrakantha, J. and Suseela, K.P. (2002). Effect of heavy metal (Cadmium borate) on the biochemical composition of *Chironomous* larvae (Diptera). *Ind. J. Environ. and Ecoplan.*, 6: 255–258.

Timmerman, K.R., Peeters, W. and Tonkes, M. (1992). Cadmium, Zinc, Lead and Copper in *Chironomus riparius* larvae (Diptera). *Hydrobiol.*, 241: 119–134.

Trin, E. (1979). *Quality Criteria for Water*. Castle House Publ. Ltd., pp. 256.

Underwood, E.J. (1971). *Trace Elements in Human and Animal Nutrition*, 3rd Ed. Academic Press, New York, pp. 208.

USEPA (1980). *Ambient and Water Quality Criteria for Cadmium*. Office of Water Regulations and Standards. Criteria and Standard Div., Washington, D. C.

USEPA (1989). *Exposure Factors Handbook*. EPA/400/8–89–043, Office of Health and Environmental Assessment, Washington, D.C.

Valek, A. (1965). Acute renal insufficiency in intoxication with mercury compounds I. Actiology clinical picture renal function. *Acad Med. Scand.*, 177: 63–67.

Verenkamp, H. (1973). Matalle in Liebensprozessen. *Chimie Unsercr Zeit.*, 7: 97–105.

Venugopal, B. and Luckey, T.D. (1975). Toxicology of non-radioactive heavy metals and their salts. In: *Heavy Metal Toxicity: Safety and Hormology*, (Eds.) Luckey, T.D., Venugopal, B. and Hutchenson, D. Stuttgert. Thieme p. 4–73.

Voege, F.A. (1971). *Levels of Mercury Contamination in Water on Hg in Man's Environment*. Ottawa Royal Society of Canada, pp. 107.

Washko, P.W. and Cousins, R.J. (1977). Role of dietary calcium and ca1cium binding protein in cadmium toxicity in rats. *J. Nutr.*, 107: 920–928.

Wasserman, G., Graziano, J.H., Factor Litrak, P., Papovae, D. and Morina, N. (1992). Independent effects of lead exposures and iron deficiency amaenia on development outcome at age 2 yrs. *J. Pediatr.*, 121: 695–703.

WHO (1973). Long term programme in environmental pollution control in Europe. The hazards to health of persistent sub. In: *Water*, Capeggen, pp. 23.

WHO (1976). *Mercury Environmental Health Criteria*. I. Geneva.

WHO (1977). *Environmental Health Criteria*: *Lead*. Geneva, Switzerland, pp. 3.

Ziegler, E.E., Edwards, B.B., Jensen, R.L., Mahaffey, R.R. and Fomon, S.J. (1978). Absorption and retention of lead by infants. *Pediatr. Res.*, 12: 29–34.

Chapter 2

Heavy Metal Pollution in Marine and Estuarine Environment: The Case of North and South 24 Parganas Districts of Coastal West Bengal

Sutapa Das, Abhijit Mitra, Kakoli Banerjee, Debarati Mukherjee and D.P. Bhattacharya

ABSTRACT

Concentrations of Zn and Cu were estimated at two different stations in North and South 24 Parganas districts of coastal West Bengal. A review of the past work was also done to evaluate the alteration in the present environmental matrix. The concept of biomagnification of conservative pollutants upto second tier was confirmed by analyzing the members of pelagic and detritus food chains in both the districts.

***Keywords**: Pelagic, Biomagnification, Detritus.*

Introduction

Marine pollution has been defined by the Intergovernmental Oceanographic Commission as the introduction by man, directly or indirectly, of substances or energy sources into the marine environment (including estuaries) which results in deleterious effects as harm to living resources, hazards to human health, hindrance to marine activities, including fishing; impairment of quality for use of sea water and reduction of amenities (www.fao.org/docrep/meeting/003/s0645e/s0645e00.htm). Some of the substances regarded as pollutants, like, heavy metals and petroleum hydrocarbons occur naturally in the sea and human introductions add to natural concentrations. Few introduced pollutants decompose with time or get diluted by the very large volume of the oceans so that their effects are not noticeable. Some of the major anthropogenic pollutants and their effects are listed in tabular form (Table 2.1).

Table 2.1: Some Major Forms of Marine Pollution and their Effects

Sl.No.	*Pollutant*	*Location*	*Effects*
1.	Petroleum hydrocarbons	Local oil spill	Mass mortality of benthos and sea birds. Effect of low level concentration is unknown.
2.	Plastics	Beaches, floating debris	Aesthetically disturbing, entanglement of animals, ingestion by marine animals.
3.	Pesticides and related compounds	Local point source inputs	Acute toxicity, long term sublethal effects largely unknown.
4.	Heavy metals	Industrial outfalls	Mostly sublethal effects, cause growth abnormalities.
5.	Sewage	Local outfalls, agricultural runoff	Eutrophication and alteration of community structure, introduction of pathogens.
6.	Radioactive wastes	Local power plants, sea dumping sites	Generally considered to be below harmful level, biomagnification reported in many cases.
7.	Thermal effects	Local power plants	Warming leads to alteration of community structure.

About 70 per cent of the earth's surface is occupied by salt water. This compartment is a unique source of food, chemicals, oil, petroleum hydrocarbons and natural gas. The diverse gene pool present in this salty aquatic phase has given this compartment a special status in the matrix of global biodiversity. However, rapid growth of population coupled with intense industrialization has posed a negative impact on the positive health of the marine and estuarine ecosystem. Basically, the major sources of marine pollution are:

1. Domestic sewage
2. Sewage sludge
3. Industrial wastes
4. Solid wastes
5. Shipboard wastes
6. Aquacultural farms
7. Pesticides and fertilizers
8. Offshore oil exploration and production wastes
9. Oil spills
10. Radioactive wastes
11. Heat: Thermal pollution from power plants
12. Flyash from thermal power plants
13. Continental run-off
14. Antifouling paints
15. Berges and other metallic structures
16. Ocean mining
17. Precipitation of air-borne pollutants
18. Oil from tanker cleaning and deballasting
19. Weathering of the earth's crust
20. Volcanic eruptions
21. Natural submarine oil seeps
22. Dredge spoils
23. Military wastes (weapon testing etc.)
24. Tourism and recreational activities.

These sources may be categorized into two broad headings:

Point Sources

Point pollution originates from a direct source like a factory outfall pipe or an aquaculture unit etc.

Non-point sources

Some sources of non-point pollution are runoff from farming (fertilizer manure), industrial runoff (heavy metals, phosphorus), urban runoff (oils, salts, various chemicals) and atmospheric fallout of air borne pollutants. Non-point pollution is very difficult to control as it originates from multiple sources.

With this background, the present article aims to monitor the levels of two conservative pollutants (Zn and Cu) in the ambient environment and members of trophic levels existing in the brackish water system of North and South 24 Parganas districts. A review of the past work was also done to analyse the magnitude of alteration in the present geographical locale.

Materials and Methods

The entire network of the present programme was divided into three phases.

Phase I: Selection of Stations

Geographically, the study area encompasses the coastal zone of two major districts in the state of West Bengal namely, 24 Parganas (North) and 24 Parganas (South). These two districts have special status due to the existence of Indian Sundarbans, which has been declared as the World Heritage Site by UNESCO in the year of 1989. Two stations, one from each districts, were selected for the present study.

Malancha (station 1) is situated in the low saline, upper stretch of Indian Sundarbans almost near to the northern boundary of Sundarban Biosphere Reserve.

Sagar Island (station 11) is situated in the high saline, lower stretch of Indian Sundarbans at the confluence of the Hooghly river and Bay of Bengal.

Phase II: Analysis of Heavy Metals in the Ambient Media and Biological Samples

Analysis of Dissolved Heavy Metals in Water

The sampling of surface water at high tides was done during 2003 in both the stations. Before analysis, each water sample was collected and stored in clean TARSON bottles and was filtered through a 0.45 µm millipore membrane. The filtrate was treated with diethyl dithiocarbamate and extracted in carbon tetrachloride (Chakroborti *et al.*, 1987). The extract was evaporated to dryness and the residue was mineralized with 0.1 ml of concentrated nitric acid. Analytical blanks were prepared and treated with the same reagents. Analyses were done in duplicate by direct aspiration into AAS (Perkin-Elmer Model 3030) equipped with a HGA-500 graphite furnace atomizer and a deuterium background corrector.

Analysis of Biologically Available Heavy Metals in Sediments (Detritus)

The sampling of detritus rich surface sediments from the crab culture sites at both the stations were done simultaneously alongwith water sampling during 2003. Each sample was first freed from visible shells or shell fragments, dried overnight at 105°C in a ventilated oven and then crushed in a porcelain mortar to pass through a 0.45 mm nylon sieve. The analytical procedure used was the dilute (0.5N HCl) acid treatment of Malo (1977) for the determination of the biologically available fraction of the total trace elements.

Analysis of Heavy Metals in Biological Samples (Plankton and Crabs)

Phytoplankton samples were collected during 2003 through a vertical tow of a plankton net (20 µm effective mesh size) at each station during high tide condition at noon. The plankton net was approximately 50 cm long, with a 26 cm diameter mouth and a 10 cm diameter opening at the cod end which was tied to a 125 ml TARSON collection bottle. These samples were collected and preserved using 1 ml of 37 per cent formaldehyde (~2 per cent final concentration). About 5 gm of phytoplankton samples were collected from each station by this method. For zooplankton collection, the effective mesh size of the net was transformed to 60 µm.

Crabs (*Scylla serrata*) were collected from both the stations from the local culturists. Muscles were carefully isolated from crab and washed with doubled distilled water. Crab muscles and plankton were oven dried at 110°C. The metal contents of the dried biological samples were estimated employing the Standard Method of digestion and subsequently using Atomic Absorption Spectrophotometer (Perkin-Elmer Model 3030) equipped with a HGA-500 graphite furnace atomizer and a deuterium background corrector (Harper *et al.*, 1989).

Phase III: Review Work

In this phase, secondary data were collected from sources like Fisheries Department and various Academic Institutes in order to carry out a time series analysis.

Results and Discussion

The coastal zone of West Bengal is one of the most biologically productive, taxonomically diverse and aesthetically celebrated zones of the Indian sub-continent, which is sustaining the famous mangrove chunk of Indian Sundarbans (Mitra *et al.*, 1992). However, due to rapid industrialization and urbanization of the city of Kolkata, Howrah and the newly developing Haldia complex, a negative impact has been exerted on the positive health of the ecosystem (Mitra and Choudhury, 1993).

The river systems and channels in and around coastal West Bengal have undergone considerable degradation. Several outlets of drainage channels of the river system such as Kultigang and Bidyadhari have deteriorated due to heavy discharge of untreated sewage from the highly urbanized city of Kolkata. Bidyadhari and Piyali rivers are practically dead. Such activities have posed an adverse impact on the Matla estuarine system. Until recently, Ichhamati, Bidyadhari, Kalagachia, Matla, Moni, Saptamukhi and Hataniaduania had plenty of fish and prawn seed resources, but pollution load, siltation and imbalance in the river ecosystem have depleted these resources (Choudhuri and Choudhury, 1994). The main reason that can be pinpointed behind the deterioration of water quality in the coastal zone of West Bengal are as follows:

1. Release of untreated wastes from the city of Kolkata, Howrah and the newly developing Haldia complex. According to UNEP report, 1125 million litre of waste water is discharged per dar through Hooghly estuary. The lower stretch of the estuary receives a waste water load of 396×10^3 km^3 per hour along with the annual runoff 493 km^3 (Fisheries Department, 2003).
2. Release of oil and grease from the adjacent landmasses, fishing vessels, trawlers, ships and various industries mainly concentrated in the lower stretch of the Hooghly estuary.
3. Entry of heavy metals (like Zn, Cu, Fe, Co, Ni, Pb etc.) in the estuarine system not only from the untreated effluents of the factories and industries, but also from the fishing vessels and trawlers that use antifouling paints for their maintenance, whose main ingredients are Zn, Cu and Pb.

Table 2.2: Year-wise Concentrations of Dissolved Heavy Metals (Zn and Cu in µg/1) in Water Samples at Different Locations

Year	Location	Zn (µg/l)			Cu (µg/l)			References
		A	B	C	A	B	C	
1987	Sagar South	73.60	103.50	90.97	3.80	8.91	7.54	Mitra and Choudhury, 1994
1988	Sagar South	82.02	105.52	94.52	5.05	8.79	8.29	Mitra and Choudhury, 1994
1989	Sagar South	84.47	NA	NA	4.43	NA	NA	Mitra and Choudhury, 1994
1992	Kulti	221.97	293.50	277.10	192.80	268.57	247.47	Mitra *et al.*, 1999
1993	Kulti	210.85	279.12	266.45	180.45	255.45	234.87	Mitra *et al.*, 1999
2002	Northern tip of Sagar Island	455.06	456.95	455.66	223.07	308.01	266.01	Das *et al.*, 2003
2003	Northern tip of Sagar Island	442.21	NA	NA	226.43	NA	NA	Das *et al.*, 2003
2003	Ghushighata (Haroa)		80			10		Fisheries Dept., 2003
2003	Sandeshkhali		30			20		Fisheries Dept., 2003
2003	Namkhana I		10			20		Fisheries Dept., 2003
2003	Namkhana II		40			10		Fisheries Dept., 2003

NA: Data not available.

A: Pre monsoon (March–June); B: Monsoon (July–October); C: Post monsoon (November–February).

Table 2.3: Year-wise Concentrations of Biologically Available Heavy Metals (Zn and Cu in mg/kg dry wt.) in Surface Sediments at Different Locations

Year	Location	Zn (mg/kg dry wt.)			Cu (mg/kg dry wt.)			References
		A	B	C	A	B	C	
1992	Kulti	216.75	167.45	191.35	90.75	52.52	66.75	Mitra *et al.*, 1999
1993	Kulti	193.67	164.37	201.70	83.00	62.87	70.60	Mitra *et al.*, 1999
2002	Northern tip of Sagar Island	66.52	62.30	86.34	54.31	34.83	36.66	Das *et al.*, 2003
2003	Northern tip of Sagar Island	78.75	NA	NA	49.23	NA	NA	Das *et al.*, 2003
2003	Ghushighata (Haroa)		6500			5700		Fisheries Dept., 2003
2003	Sandeshkhali		5600			NA		Fisheries Dept., 2003
2003	Namkhana I		5600			NA		Fisheries Dept., 2003
2003	Namkhana II		6500			5700		Fisheries Dept., 2003

NA: Data not available.

A: Pre monsoon (March–June); B: Monsoon (July–October); C: Post monsoon (November–February).

This picture of pollution has been reflected through the data bank collected through secondary survey.

The adverse effect of coastal pollution is often reflected through analysis of food chain components in an ecosystem. We analyzed a detritus food chain and a plankton based food chain in recent time to focus on the heavy metal pollution profile and evaluate the process of bioaccumulation and biomagnification.

Table 2.4: Accumulation Pattern of Heavy Metals in Members of Benthic Compartment (in µg/gm dry wt.) and Detritus (in µg/gm dry wt.) Around Malancha (North 24 Parganas) During 2003

Compartment (Benthic)	*Cu*			*Zn*		
	A	*B*	*C*	*A*	*B*	*C*
Detritus	117.56	98.65	105.33	65.98	38.96	45.32
Crab	307.23	102.04	211.65	121.56	83.65	99.56

A: Pre monsoon (March–June)

B: Monsoon (July–October)

C: Post monsoon (November–February).

Table 2.5: Accumulation Pattern of Heavy Metals in Members of Benthic Compartment (in µg/gm dry wt) and Detritus (in µg/gm dry wt.) at Sagar Island (South 24 Parganas) During 2003

Compartment (Benthic)	*Cu*			*Zn*		
	A	*B*	*C*	*A*	*B*	*C*
Detritus	91.53	84.68	89.21	51.37	35.62	47.85
Crab	213.44	96.21	191.52	81.52	53.66	70.09

A: Pre monsoon (March–June)

B: Monsoon (July–October)

C: Post monsoon (November–February).

The present investigation reflects that the concentration of Cu in both the stations is greater than Zn in detritus food chain, but in pelagic food chain, the concentration of Zn has superceded the Cu level. A significant spatial variation of metal level had been observed in the present study. The values were comparatively higher at station I (Malancha) which may be attributed to its location adjacent to the highly urbanized city of Kolkata. Here, the major sources of pollution are industrial runoff (mainly from tanneries, battery manufacturing units and several miscellaneous factories situated within and outskirts of the city of Kolkata).On the other hand, in case of station II (Sagar Island), the major sources of pollution are Haldia port-cum-industrial complex, industrial units situated in the lower stretch of the Hooghly estuary, fishing vessels and trawlers (that use antifouling paints for their conditioning) and agricultural runoff.

Table 2.6: Accumulation Pattern of Heavy Metals in Members of Pelagic Compartment (in µg/gm dry wt) and Water (in µg/l) Around Malancha (North 24 Parganas) During 2003

Compartment (Pelagic)	*Cu*			*Zn*		
	A	*B*	*C*	*A*	*B*	*C*
Water	31.42	44.20	36.20	121.48	260.02	210.79
Phytoplankton	46.37	50.02	48.11	197.85	283.42	277.20
Zooplankton	48.85	71.20	61.05	202.28	290.08	285.20
Hilsa	35.65	40.09	38.13	52.70	75.29	66.20

A: Pre monsoon (March–June)

B: Monsoon (July–October)

C: Post monsoon (November–February).

Table 2.7: Accumulation Pattern of Heavy Metals in Members of Pelagic Compartment (in µg/gm dry wt.) and Water (in µg/l) at Sagar Island (South 24 Parganas) During 2003

Compartment (Pelagic)	*Cu*			*Zn*		
	A	*B*	*C*	*A*	*B*	*C*
Water	15.29	28.44	19.65	108.52	207.85	193.79
Phytoplankton	20.28	31.75	26.30	156.39	268.29	217.84
Zooplankton	31.32	56.28	47.29	170.55	279.41	266.22
Hilsa	22.02	29.16	26.29	46.82	39.38	51.38

A: Pre monsoon (March–June)

B: Monsoon (July–October)

C: Post monsoon (November–February).

The seasonal variation of heavy metals was also pronounced in the present data set. In case of detritus food chain, both the heavy metals were found in the order, pre-monsoon > post-monsoon > monsoon. However, in case of pelagic food chain, the metals showed a dominance in monsoon followed by post-monsoon and pre-monsoon. The high concentration of heavy metals in the aquatic phase during monsoon may be due to runoff from adjacent landmasses.

The heavy metals from the aquatic phase and the detritus have been transmitted to dependent organisms of the systems. The pelagic organisms who depend on the water for food supply and the crabs, whose main diet is the detrital matter have been found to accumulate substantial level of Zn and Cu in their tissues. The phenomena of bioaccumulation and biomagnification have intensified with the concentration of heavy metals at different trophic levels (Mitra *et al.*, 2000). However, a critical examination of the data set reveals that the concentrations of Zn and Cu have not intensified at the third trophic level (from zooplankton to Hilsa fish). The accumulation of metal in different species is a function of their respective membrane permeability and enzyme system. Species which have efficient excretory mechanism are not good accumulator of pollutants and hence in such species, the concept of biomagnification becomes null and void. Hence, an indepth knowledge of biology and physiology of species is necessary to explain the process of biomagnification.

References

A Draft Report on Sundarban Wetland (2003). A report of the Department of Fisheries, Aquaculture, Aquatic Resources and Fishing Harbours, Government of West Bengal.

Chakraborti, D., Adams, F., Van Mol, W. and J.K. Irgolic (1987). Determination of trace metals in natural waters at nanogram per litre levels by Electrothermal Atomic Absorption Spectrophotometry after extraction with sodium diethyldithiocarbamate. *Anal. Chem. Acta.*, 196: 23–31.

Chaudhuri, A.B. and A. Choudhury 1994. In: *Mangroves of the Sundarbans*, Volume One: IUCN–The World Conservation Union, India.

Das, Jayeeta, Halder, K.C., Banerjee, Kakoli, Bhattacharya, D.P. and Abhijit Mitra (2003). Bioaccumulation of Zn, Cu and Pb in

Catenella repens inhabiting the Northern tip of Sagar Island, Sundarbans. *Sea Explorers*. 6: 63–68.

Harper, D.J., Fileman, C.F., May, P.V. and J.E. Postmann (1989). The analysis of trace metals in marine and other samples Aquatic Environmental Protocols: Analytical Methods. MAFF *Direct Fish Lowestoft*, UK.

Malo, B.A. (1977). Partial extraction of metals from aquatic sediments. *Environ. Sci. Technol.*, 3: 277–288.

Mitra, A. and Choudhury, A. (1994). Dissolved trace metals in surface waters around Sagar Island. *India. J. Ecobiol.*, 6(2): 135–139.

Mitra, A., Mandal, T., Jamaddar, Y.A. and D.P. Bhattacharya (1999). Inter-relationship between condition coefficient (K) of edible prawn and heavy metals from Kulti brackish water pond, West Bengal, India. *Geobios*, 26: 187–190, Vol. 26(4).

Mitra, A., Mitra, S., Hazra, S. and A. Chaudhuri (2000). Heavy metal concentration in India coastal fishes. *Res. J. Chem. Environ.*, 4(4): 35.

Mitra, Abhijit and Amalesh Choudhury (1993). Heavy metal concentrations in oyster *Crassostrea cucullata* of Sagar Island, India. *Indian Journal of Environmental Health, NEERI*, 35(2): 139–141.

Mitra, Abhijit, Amalesh Choudhury, and Zamaddar, Ali and Yusuf (1992). Seasonal variations in metal content in the gastropod *Cerithedia* (*Cerithideopsis*) *cingulata*. *Proceedings of the Zoological Society*, 45: 497–500.

www.fao.org/docrep/meeting/003/s0645e/s0645e00.htm.

Chapter 3

Heavy Metal Toxicity and Bio-accumulation: A Critical Appraisal

Bharat Bhusan Patnaik, J. Hongray Howrelia and M. Selvanayagam

ABSTRACT

Food, water and the environment have deteriorated worldwide to the point that we all are vulnerable to chronic, low level exposure to toxic metals. Slow accumulation of toxic metals have multiple symptoms and are rather non-descript. The sulfhydryl-reactive metals (Mercury, Lead, Arsenic and Cadmium) are insidious and affect biochemical, molecular and nutritional process. Basic research on their aspects has also reaveled their inhibition of anti-oxidative enzymes and depletion of intracellular glutathione. Increased surface and molecular bio-accumulation of these metals have led to the triggering of hsp gene and consequent induction of heat shock proteins. This review critically enlists the mechanisms by which toxic metals accumulate and affect the metabolic processes.

Keywords: Heavy metals, Bio-accumulation, Sulfhydryl-reactive metals, hsp.

Introduction

Heavy metals are widespread pollutants of great environmental pollutants of great environmental concern as they are non-degradable and thus persistent. They are regarded to be cytotoxic, mutagenic and carcinogenic. It is well perceived that there is a permissible limit of each metal, above which they are generally toxic and some are even hazardous.

The real problem today is not to determine if heavy metals like copper, zinc, cadmium etc. are toxic to fish and other aquatic organisms: we know they are. Hence, the aim of aquatic toxicology must be to reveal subtler and insidious changes induced by heavy metals on aquatic organisms and their environment. Such negative effects of heavy metal pollution can be studied at different levels of biological organization from responses at sub-cellular level of an individual to alteration at the population and biological community level. Often sublethal effects are biochemical in origin as most metals exert their effect at a basic level of the organism by reacting with enzyme or metabolite in enzymatic reaction, or by binding to and interacting with membrane structures or other functional components of the cell. Such primary interactions between the toxic substance and various cell components may induce a sequence of structural and functional alteration at a higher level of organization, manifested by and impairment of vital functions, such as nerve and muscle functions, respiration, circulation, immune defense, osmoregulation and hormonal regulation. Ultimately such effects might lead to irreversible and detrimental disturbances of integrated function, such as behaviors, growth, reproduction and survival, which in their turn may cause changes at the population and biological community levels. Thus the continued existence of a fish population is ultimately a function of the responses of individual organisms to the environmental changes. If responses can be recognized at sublethal levels it may be possible to detect disturbances and threats at an earlier stage, as individual responses always proceed population responses.

Chronic, low level exposure to toxic metals is an increasing global problem. The symptoms associated with the slow accumulation of toxic metals are multiple and non-descript and overt expression of toxic effects may not appear until later in life. The sulfhydryl-reactive metals (mercury, cadmium, lead, arsenic) are particularly insidious and can affect avast array of biochemical and nutritional processes.

Table 3.1: Environmental Factors and their General Effect on the Speciation and Uptake of Heavy Metals

Environmental Variable	*Modulating Factor*	*Effects*
Temperature	Metabolic rate oxygen consumption	Increased respiration rate leads to increased uptake
Water hardness	Ca^{2+}, Co^{3}	Ca^{2+} is an antagonist to Cd^{2+} and Zn^{2+} by its competition for Ca sites in pores and pumps. Carbonate forms complexes with heavy metals
pH	H^{+}, OH^{-}	Low pH leads to increased H^{+} and free metal ion concentration. High pH leads to metal-hydroxyl formation
Organic compounds	Humic substances	Complexation and decreased uptake
Complexing agents	EDTA, NTA, Xanthates	Complexation and decreased uptake
Salinity	Cl^{-}	Increased formation of metal chlorides and a decrease in uptake

Metals do vary somewhat with respect to primary sites of deposition. For *e.g.* Mercury and Cadmium are deposited heavily in the kidneys (Herber, 1994 and Nylander *et al.*, 1987).

However, unlike mercury, cadmium does not readily cross the blood-brain barrier in adults, and in contrast to mercury, cadmium is associated more with peripheral neuropathy than disorders of the central nervous system (Chang, 1996). Lead is deposited primarily in bone (Rabinowitz *et al.*, 1976), and disrupts erythropoiesis (Rabinowitz *et al.*, 1986 and Schwartz *et al.*, 1990). Zinc is primarily deposited in bone, skin and muscle tissue and to a lesser extent in liver, kidney and testis (Pentreath, 1973 and Wicklund, 1990) whereas copper accumulates in the liver (Hogstrand *et al.*, 1991).

What are the Primary Biochemical Processes Disrupted by Sulfhydryl-reactive Metals?

This question is best answered by focusing on mercury, which has been the subject of extensive basic research.

Mercury enters water as a natural process of offgassing from the earth's crust and as a result of industrial pollution. Mercury is methylated by algae and bacteria in water and moves up the food chain to highest concentration in large predatory fish such as swordship, shark, salmon and tuna. Other sources of mercury include the combustion of fossil fuels, and the production of chlorine, paper and pulp, fungicides/seed preservatives, and paints. Large amount of mercury also enter the environment as a result of careless processing of gold from ore.

The two major, highly absorbed sub-species of mercury are elemental mercury and methylmercury (MeHg). Although elemental mercury is poorly absorbed if ingested, mercury vapor is efficiently absorbed through the lungs and quickly passes the blood-brain barrier. Due to its lipophilic nature, elemental mercury has a high affinity for myelin and lipid membranes. Once inside the cell, mercury is oxidized by catalase to the highly reactive mercury ions.

Methylmercury can be demethylated and oxidized to mercury ions. Once assimilated in the cell, mercury ions and methylmercury ions form covalent bonds with glutathione and cysteine residues of proteins.

Once absorbed, mercury has a low excretion rate. A significant proportion of the assimilated mercury is retained and continually accumulates in the kidneys, neurological tissue (including the brain), and the liver. The overt neurotoxic effects, of high level mercury exposure are well established, but the subtle effects of chronic, low level mercury accumulation appear to be vast and non-descript.

The sulfhydryl-reactive metals have 3 major properties that mechanistically explain how they elicit a majority of their toxic effects. First, they are transition metals that promote the formation of hydrogen peroxide and enhance the subsequent iron and copper induced production of lipid peroxide and the highly reactive hydroxyl radical (Miller *et al.*, 1991; Halliwell and Gutteridge, 1989). Lipid peroxide alters membrane structure and is highly disruptive of mitochondrial function.

Mercury and cadmium have high affinities for glutathione (GSH), which is the primary intracellular antioxidant and conjugating agent. Importantly, a single atom of cadmium or mercury can bind to, and cause the irreversible excretion of upto two GSH tri-peptides (Zalups and Lash, 1996). The metal-GSH conjugation process is desirable in that it results in the excretion of the toxic metal into the bile. However, it can deplete the cell of GSH and thus decrease antioxidant capacity.

Lead-induced depletion of intracellular GSH and increased levels of malondialdehyde in brain and liver have been demonstrated in animal models (Gong and Evans, 1997). Mercury not only directly removes GSH from the cell, but also inhibits the activities of two key enzymes involved in GSH metabolism: GSH reductase and GSH synthatase (Zalups and Lash, 1996). Mercury also inhibits the activities of the free radical quenching enzymes catalase, superoxide dismutase (Benov *et al.*, 1990) and perhaps GSH peroxidase. The inhibition of GSH peroxidase has been attributed to the formation of mercury-selenide complex (Cuvin-Aralar and Furness, 1991).

The sulfhydryl-reactive metals disrupt the structure and function of numerous important proteins through direct binding to free sulfhydryl groups. Infact sulfhydryl groups are critical for the biological activities of virtually all proteins, including Na/K ATPase. Metal induced inhibition of Na/K ATPase can result in astrocyte swelling and destruction (Aschner *et al.*, 1990).

Mercury inhibits the polymerization of tubulin, causes depolymerization of existing microtubules, and in animal studies results in brain lesions that closely resemble those found in patients with Alzheimer's disease (Pendergrass *et al.*, 1997).

Heavy Metal Toxicity and Adaptation

The body makes important adaptive changes in response to exposure to sulfhydryl-reactive metals. Short and long-term exposure to Methylmercury in drinking water resulted in a two- to three-fold upregulation of mRNA encoding for g-glutamylcysteinesynthatase (Wood and Ellis, 1995), which is the rate-limiting enzyme in GSH synthesis.

A second adaptive and protective response to toxic metal exposure is induction of metallothionein synthesis. Metallothioneins are a fascinating group of low-molecular weight, intracellular proteins that serve as storage depot for copper and zinc, and "scavenge" sulfhydryl-reactive metals that enter the cell. Metallothioneins across species are rich in cysteine (~30 per cent) and have higher affinities for mercury and cadmium than for zinc (Hamer, 1986). Therefore, as mercury and cadmium bind to metallothionein, and are restricted from entering the mitochondria, zinc is released. The free, ionized zinc, which would be toxic if permitted to accumulate, binds to a metal regulatory element on the promoter region of the metallothionein gene and "turns on" the synthesis of metallothionein (Hamer, 1986). Such induction of metallothionein provides increased binding capacity for both toxic metals (Protective) and zinc (functional).

Environmental exposure of fish to heavy metals such as cadmium results in accumulation of the metal in the hepatic and renal metallothionein pool and induction of de novo synthesis of metallothionein (Olsson *et al.*, 1989a). While heavy metals may be toxic to animal at low doses, the binding of the metals to metallothionein most likely inhibits the toxic action of the metals. However, if an animal is exposed to high doses of metals the normally low levels of metallothionein would not be sufficient to sequester all the metal taken in and the rate of de novo synthesis of metallothionein would be too low to protect the cell. This indicates that there is a threshold of toxicity of the metals.

Evidence from the work undertaken in our laboratory has indicated the triggering of hsp70 gene activity due to increased bioaccumulation of heavy metal mercury in the brain of a teleost *Oreochromis mossambicus*. This has led us to believe that metal toxicity and bioaccumulation leads to adaptive immune response on the part of the organism in the form of heat shock proteins (Stress proteins). Once triggered, they prevent folding and refolding of the polypeptides and prevent active denaturation of important cellular proteins (Patnaik *et al.*, 2003).

Transport of Metals Across the Blood-brain Barrier

Cysteine, a conditionally essential amino acid, can be depleted with the chronic stress of metal burden. Cysteine becomes a pivotal factor to support detoxification and the body's attempt to produce more GSH and metallothionein.

L-leucine inhibits transport of the MeHg-cysteine complex across the blood-brain barrier (Aschner *et al.*, 1990; Kerper *et al.*, 1992). Cell studies indicate mercury exposure directly affects uptake and release of dopamine, nor-epinephrine and serotonin

(Komulainen and Tuomisto, 1982). Indirectly, mercury burden can be associated with depletion or poor assimilation of specific aminoacids that are precursors of neurotransmitters.

Metals and its Endocrine Involvement

The endocrine system (the master regulator of metabolism) is also affected by mercury burden. Like cadmium, mercury inhibits the conversion of thyroxine (T4) to active T3 (Barregard *et al.*, 1994). Mercury may inhibit the conversion of prohormone T4 to T3 by interfering with selenium availability.

Mercury may also interfere with progesterone metabolism without affecting serum levels of progesterone. *In vitro* studies indicate mercury binds to a free sulfhydryl group on the progesterone receptor and thereby diminish progesterone binding and cellular response (Lundholm, 1991). Mercury induced disruptions in hormone metabolism could certainly contribute to chronic fatigue, which is one of the hallmark features of mercury burden. Another possible link of metal toxicity to chronic fatigue via metal binding to the sulfhydryl-containing antioxidant, lipoic acid, making lipoic acid unavailable for its vital role in the energy-producing tricarboxylic acid (Citric acid, Krebs) cycle.

Metals Affecting Mineral Metabolism

Essential elemental metabolism is also directly affected by toxic metal burden. Mercury and cadmium readily displace zinc and copper from metallothionein, which serves as the intracellular "sink" for these essential elements. Copper and zinc are co-factors for superoxide dismutase, and copper is required for the synthesis of catecholamines. Zinc is also critical for wound healing, immune function and the metabolism of protein and nucleic acids. Mercury binds and "wastes" selenium, which is an integral constituent of free radical protection (GSH peroxidase).

Conclusion

The sulfhydryl-reactive toxic metals have no metabolic function and their accumulation in the body has serious adverse health effects. Metal burden taxes nutritional status, which impacts negatively on antioxidative and detoxification processes. Early detection and treatment of metal burden is important for successful detoxification, and optimization of nutritional status is paramount to the prevention and treatment of metal toxicity.

References

Aschner, M., *et al.* (1990. Methylmercury uptake in rat primary astrocyte cultures: The role of the neutral aminoacid transport system. *Brain Res.*, 521: 221–228.

Barregard, L., *et al.* (1994). Endocrine function in mercury exposed chloralkali workers. *Occup. Env. Med.*, 52: 536–540.

Benov, L.C., I.C. Benchev and O.H. Monovich (1990). Thiol antidotes effect on lipid peroxidation in mercury-poisoned rats. *Chem. Biol. Interact.*, 76: 321–332.

Chang, L.W. (1996). Toxico-neurology and neuropathology induced by metals. In: *Toxicology of Metals.* (Ed.) L.W. Chang. CRC Press, Boca Raton. p. 511–535.

Cuvin-Aralar, M.L. and R.W. Furness (1991). Mercury and selenium interaction: A review. *Ecotoxicol. Env. Saf.*, 21: 348–364.

Gong, Z. and H.L. Evans (1997). Effect of chelation with meso-dimercaptosuccinic acid (DMSA) before and after the appearance of lead-induced neurotoxicity in the rat. *Toxicol. Appl. Pharmacol.*, 144: 205–214.

Halliwell, B. and J.M.C. Gutteridge (1989). *Free Radicals in Biology and Medicine.* Oxford, UK. Claredon Press.

Hamer, D.H. (1986). Metallothionein. *Annu. Rev. Biochem.*, 55: 913–951.

Herber, R.F.M. (1994). Cadmium. In: *Handbook on Metals in Clinical and Analytical Chemistry*, (Eds.) H.G. Seiber, A. Sigel and H. Sigel. Marcel Dekker Inc., New York, p. 283–297.

Hogstrand, C., G. Lithner and C. Haux (1991). The importance of metallothionein for the accumulation of copper, zinc and cadmium in environmentally exposed perch *Perca fluviatilis. Pharmacol. Toxicol.*, 68: 492–501.

Kerper, L.E., N. Ballatori and T.W. Clarkson (1992). Methylmercury transport across the blood brain barrier by an aminoacid carrier. *Am. J. Physiol.*, 262: R761–R765.

Komulainen, H. and J. Tuomisto (1982). Effects of heavy metals on monoamine uptake and release in brain synaptosomes and blood platelets. *Neurobehav. Toxicol. Teratol.*, 4: 647–649.

Lundholm, C.E. (1991). Influence of chlorinated hydrocarbons, Hg^{2+} and methylmercury^{+} on steroid hormone receptors from egg shell mucosa of domestic fowls and ducks. *Arch. Toxicol.*, 65: 220–227.

Miller, O.M., B.O. Lund and J.S. Woods (1991). Reactivity of Hg (II) with superoxide: Evidence for the catalytic dismutation of superoxide by Hg (II). *J. Biochem. Toxicol.*, 6: 293–298.

Nylander, M., L. Friberg and B. Lind (1987). Mercury concentrations in the human brain and kidneys in relation to exposure from dental amalgam fillings. *Swed. Dent. J.*, 11: 179–187.

Olsson, P.E. *et al.* (1989a). Induction of metallothionein synthesis in rainbow trout, *Salmo gairdneri*, during long term exposure to water-borne cadmium. *Fish Physiol. Biochem.*, 6: 221–229.

Patnaik, B.B., H.J. Howrelia and M. Selvanayagam (2003). Recent advances in protein folding machineries. *Bioinform. India*, 1: 40–43.

Pendergrass, J.C., B.E. Haley and M.J. Vimy (1997). Mercury vapor inhalation inhibits binding of GTP to tubulin in rat brain: Similarity to a molecular lesion in Alzheimer's diseased brain. *Neurotoxicol.*, 18: 315–324.

Pentreath, R.J. (1973). The accumulation and retention of ^{65}Zn and ^{54}Mn by the plaice, *Pleuronectus platessa* L. *J. Expt. Mar. Biol. Ecol.*, 12: 1–18.

Rabinowitz, M.B., G.W. Wetherill and J.D. Kopple (1976). Kinetic analysis of lead metabolism in healthy humans. *J. Clin. Invest.*, 58: 260–270.

Rabinowitz, M.B., A. Leviton and H. Needleman (1986). Occurrence of elevated protoporphyrin levels in relation to lead burden in infants. *Env. Res.*, 39: 253–257.

Schwartz, J., P.J. Landrigan and E.L. Baker (1990). Lead induced anemia: Dose-response relationships and evidence for a threshold. *Am. J. Pub. Hlth.*, 80: 165–168.

Wicklund, A. (1990). Metabolism of cadmium and zinc in fish. *Ph.D. Thesis*. University of Uppsala, Sweden

Woods, J.S. and M.E. Ellis (1995). Upregulation of glutathione synthesis in rat kidney by methylmercury. Relationship to mercury-induced oxidative stress. *Biochem. Pharmacol.*, 50: 1719–1724.

Zalups, R.K. and L.H. Lash (1996). Interactions between glutathione and mercury in the kidney, liver and blood. In: *Toxicology of Metals*, (Ed.) L.W. Chang. CRC Press, Boca Raton, p. 145–163.

Chapter 4

Heavy Metal Accumulation in Seaweeds in Polluted and Unpolluted Estuaries

K. Sasikumar and A.N. Subramaniyan

ABSTRACT

Three seaweeds in the Vellar and Uppanar estuaries when analysed for heavy metals, showed that Cu, Zn, Cd and Pb varied in them with their habitats. However, Chaetomorpha aerea and C. linum did not show much variations between the two habitats. Entromorpha intestinalis exhibiting low levels of Cu, Zn, Cd and Pb, could be used as a pollution indicator. Among the four metals, concentration of Zn was high in all the seaweeds followed by Pb, Cd and Cu.

Keywords: *SIPCOT industries, Heavy metals, Seaweeds, Bioaccumulation.*

Introduction

In India, seaweeds are used as raw materials for the production of agar, alginate and seaweed liquid fertilizer. Unfortunately, several toxic pollutants, unknown to the biota early, have been introduced in large quantities into the aquatic environment. Metals are natural components of seaweeds and as such are harmless to marine life, but they can build up to high concentration as a result of man's activities, as in mine tailing or industrial effluents and may then cause a risk to human consuming seafood. Group of Expert of Scientific Aspects of Marine Pollution (GESAMP, 1974) has compiled brief reviews on a number of elements especially cadmium, lead, copper and zinc which have been considered as potential pollutants. Heavy metals may accumulate, unnoticed, to toxic levels. Thus problems associated with heavy metal contamination were first highlighted in the industrially advanced countries because of their large industrial discharges and especially by incidents of mercury and cadmium pollution in Sweden and Japan (Kurland *et al.* 1960; Nitta, 1972; Goldberg, 1976).

Some heavy metals such as Cu, Zn, Cd and Pb are essential for the growth and well being of living organisms including man. However, they are likely to show toxic effects when organisms are exposed to levels higher than normally required. Other elements such as Pb and Cd are not essential for metabolic activities and exhibit toxic properties. Heavy metals form the major group of toxic pollutants among the different pollutants, as these metals tamper the harmony of the ecosystem. Most of the organisms have the ability to bioaccumulate metals from the surrounding water into their bodies.

The concentration of metals in different species living in the same environment may vary as some species contain high, whereas others contain small quantities. These variations are due to the differences in assimilation and excretion capacities, or both, as the concentration of a metal in an animal is being determined by the balance between these two processes. Studies on the concentration of heavy metals in the polluted environment were reported by Badsha and Goldspink (1982), Jaffer and Ashraf (1988), Ashraf *et al.* (1991) indicating the heavy metals to concentrate in the different compartments of aquatic ecosystem. The enriched concentration of heavy metals in aquatic ecosystem due to acidification is reported earlier by Almmer and Dickson (1974) and Campbell and Stokes (1985). The physicochemical form of metals are affected by acidification and these changes influence the bioavailability and toxicity to many organisms (William and Campbell, 1981).

Hence, in the present investigation, an attempt has been made to study the physico-chemical and heavy metal accumulation such as Cu, Zn, Cd and Pb of seaweeds occurring in the Vellar estuary which is unpolluted and the Uppanar estuary polluted from coconut husk retting grounds and effluent discharges from SIPCOT industrial complex (Small Industries Production Council of Tamil Nadu).

Materials and Methods

The Vellar estuary (Station 1; lat. 11° 20′N, long. 79° 46′E) is a small estuary, blooming with fishing and aquaculture activities (Figure 4.1). Uppanar estuary (Station 2; lat. 11° 59′N, long. 79°50 E) is also a small estuary but polluted with industrial effluents of SIPCOT industrial estate, Cuddalore. For studying the distribution, algae were collected seasonally throughout the year from October 2001 to September 2002. Meteorological data such as monthly rainfall and humidity were obtained from the Meteorological Department, Parangipettai and Cuddalore. Physical parameters such as

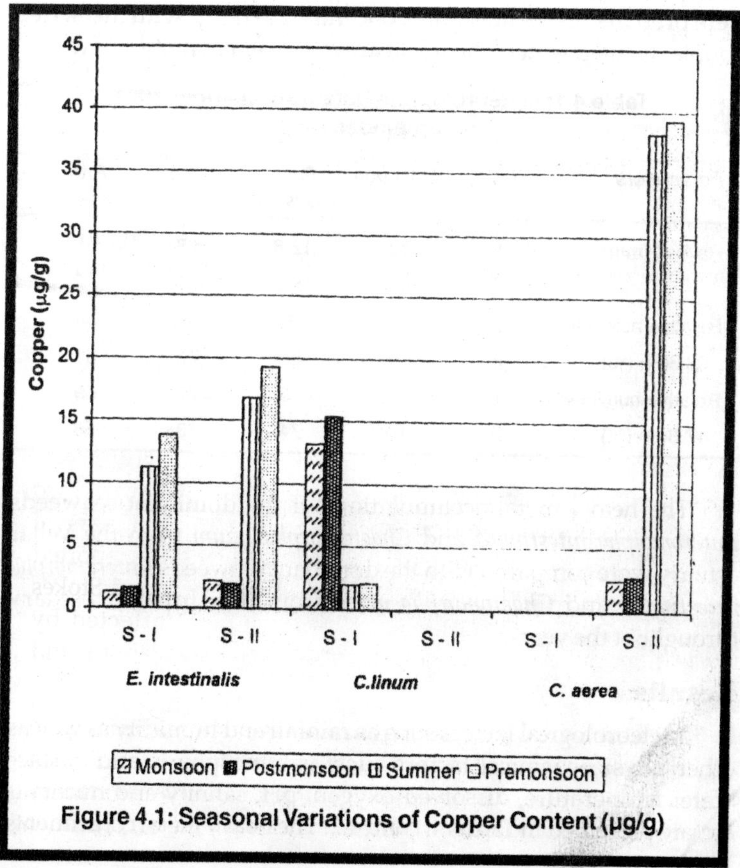

Figure 4.1: Seasonal Variations of Copper Content (µg/g)

atmospheric temperature, surface water temperature and pH were determined in the field itself. Chemical parameters such as salinity, dissolved oxygen and nutrients were studied using standard methods (Grasshoff *et al.*, 1983) in the laboratory. All these samples were dried at 60°C for 48 h and made into a fine powder. Heavy metal concentration in the samples was determined through Atomic Absorption Spectroscopy (AAS). Solvent extraction procedure was adopted for preconcentration of heavy metals from water as a preliminary step to their determination by AAS. Results are expressed in mg/l. Tissue digestion of the dried, powdered samples of different organs was carried out by ashing the samples in a muffle furnace at 550°C for two hours. The ash is further digested in conc. HNO_3, perchloric acid and sulphuric acid and dissolved in 1N HNO_3. A series of standard metal solutions (Cu, Zn, Cd and Pb) were prepared in the optimum concentration range with metal free double distilled water. The result are expressed in µg/g dry wt.

Table 4.1: Meteorological Data from October 2001 to September 2002

Parameters	*Station*	*Monsoon*	*Post-monsoon*	*Summer*	*Pre-monsoon*
Rainfall (mm)	I	333.5	12.8	5.8	8.8
	II	297.7	10.8	4.6	7.2
Relative humidity	I	64	90	85	88
08.30 h (%)	II	58	84	77	81
Relative humidity	I	80	76	77	69
17.30 h (%)	II	67	74	73	58

The heavy metal accumulations of the dominant seaweeds *Enteromolpha intestinalis* and *Chaetomorpha linum* from the Vellar estuary were compared with the dominant seaweeds *Enteromorpha intestinalis* and *Chaetomorpha aerea* from the Uppanar estuary throughout the year.

Results

Meteorological factors such as rainfall and humidity as well as other physicochemical factors such as atmospheric and surface water temperature, dissolved oxygen, pH, salinity and nutrients etc. are presented in Tables 4.1 and 4.2. The heavy metal constituents such as Copper, Zinc, Cadmium and Lead were analysed and the seasonal data are presented in Figures 4.1–4.4, respectively.

Table 4.2: Data Collected on Physico-chemical Parameters from October 2001 to September 2002

Parameters	*Station*	*Monsoon*	*Post-monsoon*	*Summer*	*Pre-monsoon*
Atmospheric	Station I	26	28	34	30
Temperature (°C)	Station II	25.5	26.4	32.6	31.2
Surface water	Station I	22	23	30	28
Temperature (°C)	Station II	21	22	29	27
Dissolved Oxygen	Station I	6.6	5.8	4.2	3.6
(ml/l)	Station II	4.8	5.9	3.7	3.9
pH	Station I	7.9	7.8	8.4	8.1
	Station II	8.0	7.9	8.25	8.18
Salinity (%)	Station I	24.8	26.1	33.4	31.6
	Station II	20.5	22.6	30.8	28.4
Phosphate (mg/l)	Station I	0.182	0.006	0.009	0.174
	Station II	1.04	0.054	0.062	1.01
Silicate (mg/l)	Station I	6.56	5.48	1.54	1.42
	Station II	7.74	6.62	1.44	1.56
Nitrate (mg/l)	Station I	0.420	0.2600	0.340	0.270
	Station II	0.451	0.290	0.438	0.248
Nitrite (mg/l)	Station I	1.057	1.042	0.378	0.392
	Station II	1.081	1.048	0.079	0.082

Copper

At Station 1, Cu concentration in *E. intestinalis* varied from 1.65 to 19.8 µg/g. Minimum (1.65 µg/g) and maximum concentrations (19.8 µg/g) were recorded during the post- and premonsoon seasons, respectively. Cu concentration in *Chaetomorpha linum* varied from 2.30 to 16.2 µg/g. Maximum concentration (2.30 µg/g) was recorded during the premonsoon season and maximum concentration (16.2 µg/g) was recorded during the postmonsoon season (Figure 4.1). *E. intestinalis* accumulated more Cu than *Chaetomorpha linum.*

At Station 2, Cu concentration in *E. intestinalis* varied from 2.40 to 16.3 µg/g. Minimum concentration (2.40 µg/g) was recorded during the postmonsoon season. Maximum concentration (16.3 μg/g) was recorded during the premonsoon season. Cu concentration in *Chaetomorpha aerea* varied from 2.50 to 39.30 μg/g. Minimum concentration (2.50 μg/g) was recorded during the monsoon season. Maximum concentration (39.30 μg/g) was recorded during the postmonsoon season (Figure 4.1). C. *aerea* accumulated more amount of Cu than *E. intestinalis.*

Zinc

At Station 1, Zn concentration in *E. intestinalis* varied from 21.40 to 130 μg/g. Minimum concentration 21.4 μg/g was recorded during

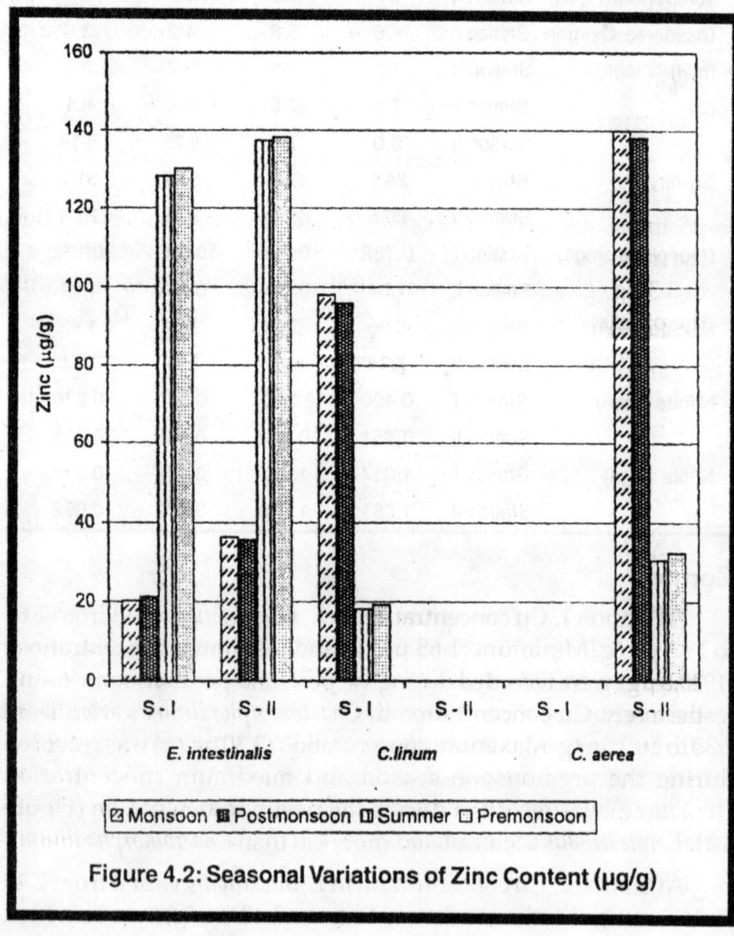

Figure 4.2: Seasonal Variations of Zinc Content (μg/g)

the postmonsoon season. Maximum concentration (130 μg/g) was noted during the premonsoon season. Zn concentration in *Cheatomorpha linum* varied from 19.3 to 98 μg/g. Minimum concentration (19.3 μg/g) was recorded during the premonsoon season. Maximum concentration (98 μg/g) was recorded during the monsoon season (Figure 4.2). *E. intestinalis* accumulated more amount of Zn than C. *linum*.

At Station 2, Zn concentration in *E. intestinalis* varied from 35.6 to 138 μg/g. Minimum concentration was recorded during the postmonsoon season, and the maximum during the premonsoon season. Zn concentration in C. *aerea* varied from 30.6 to 140 μg/g. Minimum concentration (30.6 μg/g) was recorded during the summer season. C. *aerea* accumulated more amount of Zn than *E. intestinalis* (Figure 4.2.).

Cadmium

At Station 1, Cd concentration in *E. intestinalis* varied from 2.50 to 4.8 μg/g. Minimum concentration was recorded during the postmonsoon season. Maximum concentration was recorded during the monsoon season. Cd concentration in C. *linum* varied from 1.65 to 3.3 μg/g. Both minimum and maximum were recorded during the premonsoon season (Figure 4.3).

At Station 2, Cd concentration in *E. intestinalis* varied from 2.5 to 5.6 μg/g. Minimum concentration (2.5 μg/g) was recorded during the postmonsoon season. Maximum concentration (5.6 μg/g) was noted during the monsoon season. Cd concentration in C. *aerea* varied from 1.8 to 4.5 μg/g. Minimum concentration (1.8 μg/g) was recorded during summer and maximum concentration (4.5 μg/g) during the postmonsoon season (Figure 4.3).

Lead

At Station 1, Pb concentration in *E. intestinalis* varied from 7.5 to 20.8 μg/g. Minimum concentration (7.5 μg/g) was recorded during the monsoon season. Maximum concentration (20.8 μg/g) was noted during the premonsoon season. Pb concentration in C. *linum* varied from 5.8 to 16.8 μg/g. Minimum concentration (5.8 μg/g) was recorded during the monsoon season and maximum concentration (16.8 μg/g) during the summer season (Figure 4.4). Pb concentration was higher in *E. intestinalis* than in C. *linum*.

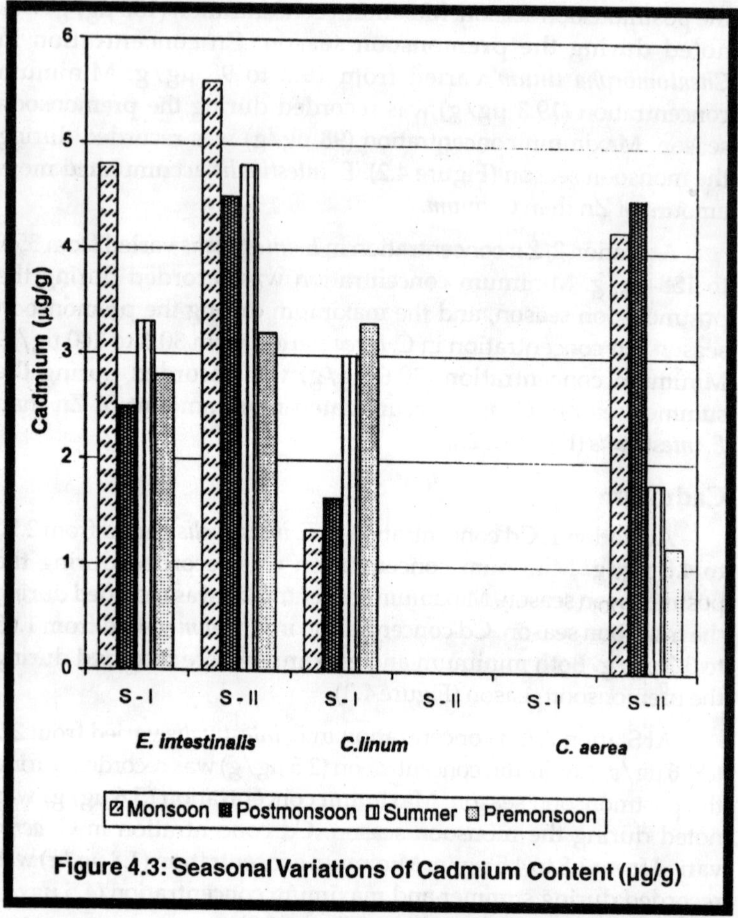

Figure 4.3: Seasonal Variations of Cadmium Content (µg/g)

At Station 2, Pb concentration in *E. intestinalis* varied from 9.8 to 25.4 µg/g. Minimum concentration (9.8 µg/g) was recorded during the monsoon season and maximum concentration (25.4 µg/g) during the premonsoon season. Pb concentration in *C. aerea* varied from 11.20 to 35.40 µg/g. Minimum concentration (11.20 µg/g) was recorded during the summer season and maximum concentration (35.40 µg/g) during the monsoon season (Figure 4.4). Pb concentration was higher in *C. aerea* than *E. intestinalis*.

Discussion

The concentration of heavy metals in the seaweed sample reveals are higher at Station 2 in the polluted area (Figures 4.1–4.4)

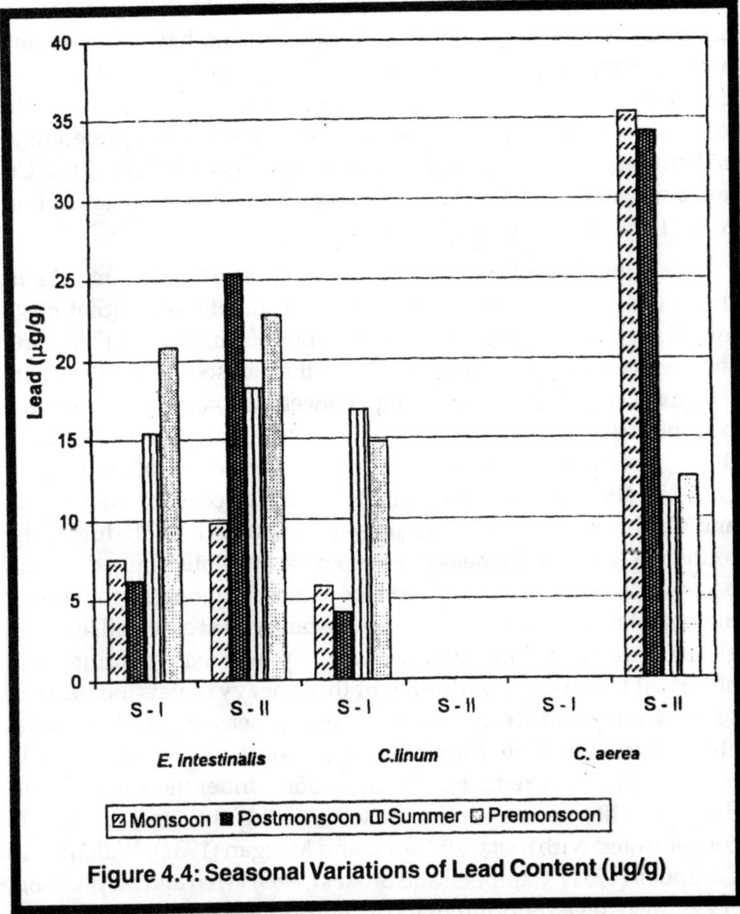

Figure 4.4: Seasonal Variations of Lead Content (µg/g)

compared to Station 1 the non-polluted area. Sources of heavy metals to Station 2 are from the discharge of industrial and domestic wastes in addition to various harbour activities. Moreover, differences in metal accumulation of seaweeds might be related not only to different metals levels in water but also to different ecological conditions such as rainfall, temperature salinity, dissolved nutrients and different geological structures (Munda and Hudnik, 1991).

The decreasing order of preferential accumulation of metals in *E. intestinalis* is Zn > Pb > Cu > Cd. Similar trend has been observed by Munda and Hudnik (1991) at Rovinj in *Enteromorpha* sp., Sivalingam (1978) at Penang in *Entermorpha flexuosa* and by

Rajendran *et al.* (1993) at Kasimedu in *E. intestinalis*. A different trend has been reported by Guvan *et al.* (1992), who have showed the order of preferential accumulation of metals in *Hepnea valentiae* as Zn > Cu > Pb > Cd. Similar trend has been observed by Kesava Rao and Indusekhar (1986) who have recorded the order of preferential accumulation of metals in *Padina boergeseni* is Zn > Pb > Cu > Cd and a similar trend has been observed by Sheila Devi Ramachandran *et al.* (1994) at Malaysia in *Padina* spp.

Seasonal variation in the distribution of heavy metals in the seaweeds shows that of the 2 species, *E. intestinalis* (Stations 1 and 2) accumulates higher concentrations of Cu, Zn, and Pb during the premonsoon season and Cd during the monsoon season. *C. linum* (Station 1) exhibits recorded high concentration of Cu, during the postmonsoon season, Cd and Pb during the summer season and Zn during the monsoon season. This trend is similar to that reported by Ganesan (1992). *C. aerea* (Station 2) shows high concentration of Cu, and Cd during the postmonsoon season and Zn and Pb during the monsoon season. The heavy metal levels are statistically different between seasons, but not between the stations. Among the four heavy metals taken for the study, zinc is substantially at elevated levels in all the seasons and in water followed by lead and cadmium. It is observed that the concentration of these heavy metals increases in premonsoon and decreases in monsoon season. The industrial effluents released into the stream at Station 2 are perhaps acidic in nature, ultimately reduce in the monsoon. Under these conditions, the solubility of metals increases making the stream heavily concentrated with metals (Stumm and Morgan (1981), William and Campbell (1981). Campbell and Stokes (1985) Wren and Stephenson, (1991) also have reported that concentration of lead and cadmium in water depends upon the acidity of surrounding medium.

Acknowledgement

Our sincere thanks are due to K. Kathiresan, CAS in Marine Biology, Annamalai University, Parangipettai and Prof. L. Kannan, Research Director for their suggestion and constant encouragement in the present work. Our thanks are also due to the authorities of Annamalai University for the facilities provided for the present investigation. We are thankful to the publishing authority for their full co-operation for the successful completion of this article.

References

Almmer, B., Dickson, W., Ekstrom, C., Homstrom, E. and U. Miler (1974). Effect of acidification on Swedish lakes. *Ambio.*, 3: 30–36.

Ashraf, M., Tariq, J. and M. Jaffer (1991). Contents of trace metals in fish, sediment and water from three freshwater reservoirs in the Indus river. *Pakistan Fisheries Research*, 12(4).

Badsha, K.S. and C.R. Goldspink (1982). Preliminary observations on the heavy metal content of four species of freshwater fish in NW England. *J. Fish Biol.*, 21 : 251–267.

Campbell, P.O. and P.M. Stokes (1985). Acidification and toxicity of metals to aquatic biota. *Can. J. Fish Aqat Sci.*, 42: 2034–2049.

Ganesan, M. (1992). Eco-biology of seaweeds of the Gulf of Manner with special reference to hydrography and heavy metals. *Ph.D. Thesis*, Annamalai University, India, pp. 162.

Gesamp. IMO/FAO UNESCO/WMO/WHO/IAEA/UN/UNEP (1974). Joint Group of Experts on the Scientific Aspects of Marine Pollution, Review of harmful substances, Supplement to the Report of Sixth session, Geneva, 22–28 March 1974, Geneva, World Health Organization.

Goldberg, E.D. (1976). The Health of the Oceans. Paris, *UNESCO*, pp. 172.

Grasshoff, K., Ehrhadt, M. and K. Kremling (1983). *Method of Sea Water Analysis*, Verlag demic gymb, Weinheim, pp. 419.

Guvan, K.C., Topcuoglu, S., Kut, D., Esen, N., Erenturk, N., Saygi, N. and E. Cevher (1992). Govener, Band Ozturk, B. Metal uptake by Black sea algae. *Bot. Mar.*, 35: 337–340.

Jaffer, M. and Ashraf, M. (1998). Selected trace metal concentrations in different tissues of fish from coastal waters of Pakistan (Arabian Sea). *Indian J. Mar. Sci.*, 17: 231–234.

Kesava Rao, C.H. and V.K. Indusekhar (1987). Chromium, lead and cadmium contents of certain seaweeds from Saurashtra coast. *Phykos*. 26: 1–7.

Kurland, L.T., Faro, S.W. and H. Siedler (1960). Minamata disease: The out-break of a neurological disorder in Minamata, Japan and its relation to ingestion of seafood containing mercury compounds. *World Neurol.*, 1: 370–395.

Munda, I.M. and V. Hudnik (1991). Trace metal content in some seaweeds from the northern Adriatic. *Bot. Mar.*, 34: 241–249.

Nitta, T. (1972). Marine pollution in Japan. In: *Marine Pollution and Sea Life* (Ed.) M. Ruivo. West Byfleet Survey, Fishing News Books, p. 77–81

Rajendran, K., Sampathkumar, P., Govindasamy, C., Ganesan, M., Kannan, R. and L. Kannan (1993). Levels of trace metals (Mn, Fe, Cu and Zn) in some Indian seaweeds. *Baseline*, 26: 283–285.

Sheila Devi Ramachandra., Phang siew Moi. and Soo Loong Tong (1994). Heavy metals content of some Malaysian seaweeds. *Proceedings of the First Asia-Pacific Conference on Algal Biotechnology*, Phang (Eds.), University of Malaya, p. 339–343.

Sivalingam, P.M. (1978). Biodeposited trace metals and mineral content studies of some tropical marine algae. *Bot. Mar.*, 21: 327–330.

Stumm, W. and J.J.Morgan (1981). *Aquatic Chemistry: An Introduction Emphasising Chemical Equilibria in Natural Waters*. Wiley-Interscience, New York, pp. 583.

William, O.N. and P.G.C. Campbell (1981). The effects of acidification on the geochemistry of Al, Cd, Pb and Hg in freshwater environments: A literature, review. *Environ. Pollut.*, 71: 91–130.

Wren, C.D. and G.L. Stephenson (1991). The effect of acidification on the accumulation and toxicity of metals to freshwater invertebrates. *Environ. Pollut.*, 71: 2–4.

Chapter 5

Heavy Metal Levels in Freshwater Bivalve Mollusc, *Lamellidens corrianus* from Nandrabad Swam Near Aurangabad

S.S. Patil

ABSTRACT

The water samples, sediments and adult bivalves, Lamellidens corrianus were collected from a fixed location from Nandrabad swamp near Aurangabad city (20 km. away) to determine heavy metals like mercury, cadmium and lead along with physiological characteristics of swamp water. In summer, temperature, pH, hardness and alkalinity were at high levels and carbondioxide and chlorides were at low levels. In monsoon chlorides and sulphates were at high and pH was at low levels while in winter, dissolved oxygen and carbon dioxide were at high levels but temperature, hardness, alkalinity and sulphates were at low levels. In summer, mercury level was high in mantle and hepatopancreas while in monsoon the level was increased in gonad and mantle but in winter it was increased in gonad, foot and adductor muscles. Cadmium level in summer was high in mantle, gills and gonad. In monsoon it was high in adductor muscles, gonad, foot and gill while in winter it was increased in adductor muscles, foot and gonad. Lead level was high in summer in foot, mantle and adductor muscles. In monsoon it was high in mantle, foot and adductor muscles but in winter it was increased in mantle ând adductor muscles than the other body parts. The levels of mercury, cadmium and lead in the swamp water, sediment and bivalve molluscs are superimposed by different seasons due to changes in environmental parameters and the changes in physiological activities of tile bivalves despite tile multiple human activities as well as domestic sewage flow and increased industrial and agricultural practices.

Keywords: *Heavy metal levels, Lamellidens corrianus.*

Introduction

The high rate of increase in human population of India and rapid pace of industrialization has created problems of disposal of domestic and industrial wastes, agrochemicals, heavy metals etc. They exert on over growing influence on near shore, bank of these water bodies and the organisms contained therein. The ability of biological organisms to concentrate the levels of dissolved undesirable substances *e.g.* lead, mercury, cadmium, copper, iron, zinc etc. and their transmission through the biological food chain is a situation for real concern. Although the complete mechanism of concentration and the transmission and their subsequent biological effects are unknown, some mitigating action is necessary (Luten *et al.*, 1986; Gangon and Fisher, 1997). The principle area concerned today is the need for understanding the levels of heavy metals and their biological effects on animal resources of economical importance. Attention on bivalve molluscs culture is fast increasing due to its commercial and nutritional value (Rao and Balaji, 1991; Patil and Mane, 2000).

The bivalve molluscs are known to be effective concentrators of trace elements and for many years these animals have been used as all indicator of contamination (Elder and Collins, 1991; Patil, 1993; Patil and Mane, 1998). It is becoming important to have broad understanding of metal levels in large variety of bivalve molluscs to establish a base line for comparison with possible future contamination in aquatic ecosystems (Dehadrai, 1991). Amongst the metal pollution in Maharashtra state, the use of cadmium, lead and mercury and other heavy metals are increasing day by day and there is a paucity of information on the heavy metal levels, its movement in food chain and web and their bioaccumulation and seasonal variations in heavy metal content from the bottom feeding animals like bivalve molluscs. The present study is aimed to carry out the base levels of mercury, cadmium and lead from water sediments and in freshwater bivalve molluscs, *Lamellidens corrianus* from Nandrabad swamp near Aurangabad city (20 km away).

Materials and Methods

The water samples, sediments and the adult bivalves, *Lamellidens corrianus* (Lee) from a fixed location were collected from Nandrabar swamp, 20 km. away from Aurangabad city, for the determination of base levels of heavy metals along with physicochemical characteristics of swamp water. The heavy metals like lead, cadmium and mercury are determined using Atomic absorption spectrophotometer, CHIMITO-200 according to standard methods of APHA (1985) and Hatch and Ott (1968) from swamp water, sediments and from different body parts of the bivalve molluscs, *Lamellidens corrianus* in summer, monsoon and winter season. Four replicates of samples were carried in every season and mean values are subjected for statistical analysis.

Results and Discussion

In summer, compared to monsoon and winter temperature, pH, hardness and alkalinity were at high levels, where as carbon dioxide and chlorides were at low levels. In monsoon chlorides and sulphates were at high levels while pH was at low levels compared to winter and summer. On the other hand, in winter dissolved oxygen and carbon dioxide were at high levels but temperature, hardness, alkalinity, and sulphates were at low levels when compared to summer and monsoon (Table 5.1).

Mercury, cadmium and lead levels in water and sediments were high in monsoon than in summer and winter (Figure 5.1). In summer mercury levels from mantle, gill and hepatopancreas of *L. corrianus* was higher than gonad, foot and adductor muscles. In monsoon, gonad and mantle showed higher level of mercury than gill, hepatopancreas, foot and adductor muscles. On the other hand in winter, gonad, foot and adductor muscles showed higher level than mantle, hepatopancreas and gill. Cadmium level in summer was high in mantle, gill and gonad than hepatopancreas, foot and adductor muscles. In monsoon it was high in adductor muscles, gonad, foot and gill than in hepatopancreas and mantle while in winter it was high in adductor muscles, foot and gonad than in hepatopancreas, gill and mantle. Lead level in summer was high in foot, mantle and adductor muscles than the other body parts. In monsoon it was high in mantle, foot and adductor muscles than in hepatopancreas, gonad and gill while in winter it was higher in mantle and adductor muscles than the other body parts (Figure 5.2).

Table 5.1: Fluctuations in Environmental Parameters on the Habitat of *Lamellidens corrianus* in Different Seasons from Nandrabad Swamp

Sl.No.	*Parameters*	*Season*		
		Summer	*Monsoon*	*Winter*
1.	Temperature (°C)	28.1–29.0	25.4–27.9	14.0–17.0
2.	pH	8.5–8.9	7.4–7.9	8.3–8.7
3.	Dissolved oxygen	3.71–4.08	5.02–5.81	6.2–6.9
4.	Carbon dioxide	3.2–8.0	5.6–8.4	8.4–9.8
5.	Hardness	268–272	158–162	140–146
6.	Chlorides	4.8–7.9	20.21–24.09	13.83–14.03
7.	Total alkalinity	57.0–71.0	43.0–59.6	28.3–34.1
8.	Sulphates	46.0–47.2	57.1–60.0	27.2–38.9

All the values in mg/lit. except temperature and pH.

Metal uptake by muscle may be from the external medium in a suitable form *i.e.* ions, chelate or complexes (Pringle *et al.*, 1986). The feeding particles found in the surroundings may influence the metal uptake (Romerill, 1971; Riley and Roth, 1974; Wilcock *et al.*, 1993). Thermodynamic calculations suggest that the Hg-chloro complexes might be responsible for increased mercury bioaccumulation. Increased industrial activities, auto exhausts, effluent from industries as well as solid waste dumping have become the sources of large quantities of heavy metals into the environment (Singh and Steinnes, 1994).

In Nandrabad swamp, multiple human and domestic animals activities as well as domestic sewage flow might have contributed mercury, cadmium and lead depositions in water and sediments. Seasonal fluctuations in the distribution of trace metals in bivalve molluscs controlled by an array of extrinsic and intrinsic factors such as extent of pollutant delivery into the water and the associated dilution in *in situ* changes in weight of organism and the direct effect of temperature, chlorinity and other water quality parameters

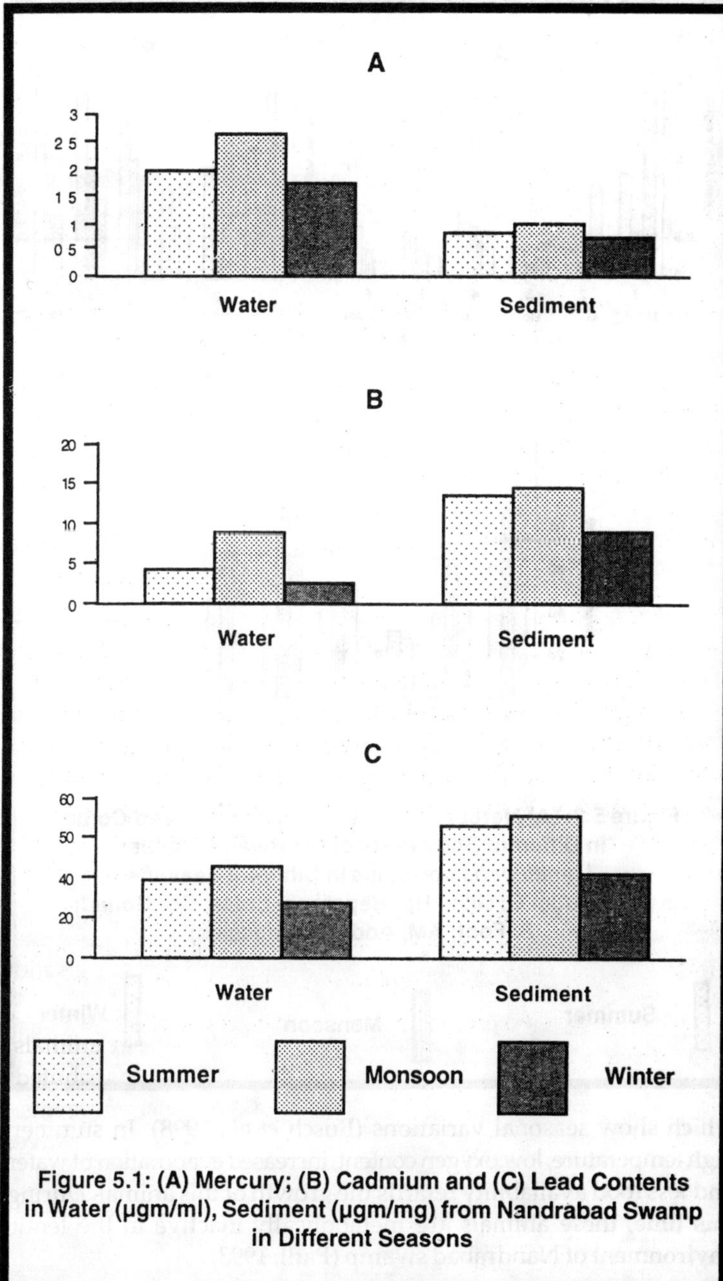

Figure 5.1: (A) Mercury; (B) Cadmium and (C) Lead Contents in Water (µgm/ml), Sediment (µgm/mg) from Nandrabad Swamp in Different Seasons

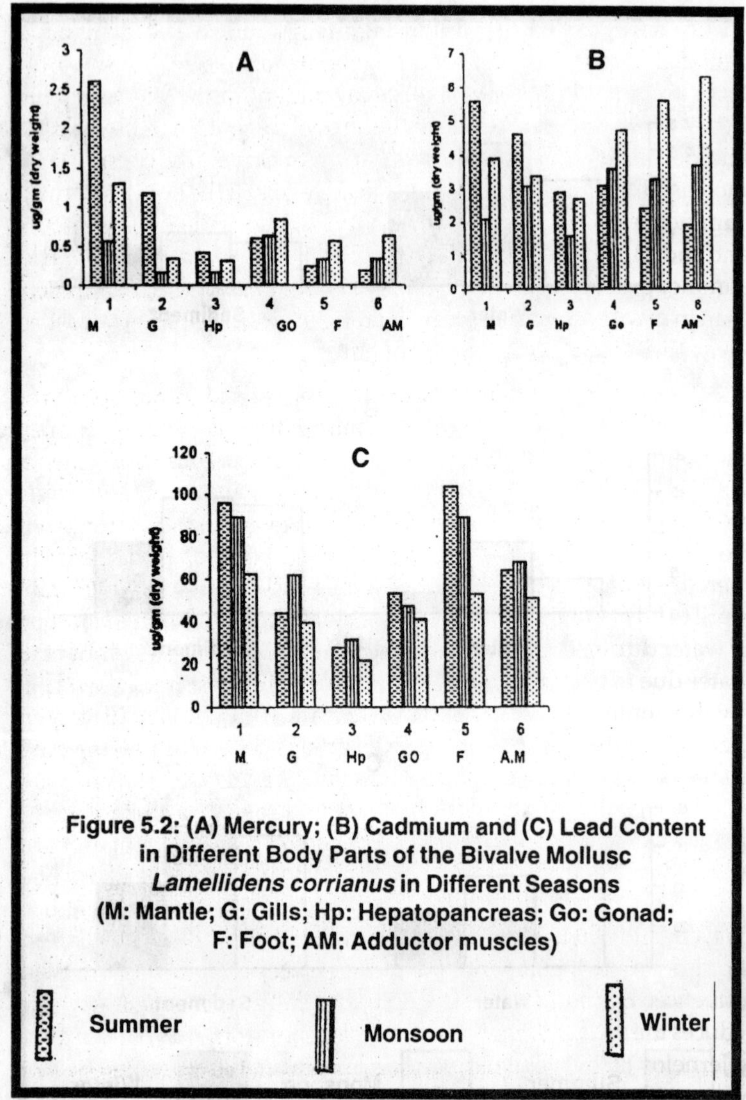

Figure 5.2: (A) Mercury; (B) Cadmium and (C) Lead Content in Different Body Parts of the Bivalve Mollusc *Lamellidens corrianus* in Different Seasons (M: Mantle; G: Gills; Hp: Hepatopancreas; Go: Gonad; F: Foot; AM: Adductor muscles)

which show seasonal variations (Busch *et al.*, 1998). In summer, high temperature, low oxygen content, increased evaporation of water and less food availability retards the growth of this animal. During this time, these animals are metabolically inactive in the lentic environment of Nandrabad swamp (Patil, 1993).

In summer the prevailing high temperature and demand to utilize more oxygen and food for maintenance metabolism could have accounted high levels of metal content in the above organs. Influx of regular sewage discharge, growth of much detritus matters and decrease in the level of swamp water could also contribute to increased mercury, cadmium and lead levels in the environment in summer season (Elder and Collins, 1991). The diffusion of mercury and cadmium through pallial edges and gills of *L. corrianus* during summer probably gave high levels of mercury and cadmium content than in monsoon. In monsoon, flooding in swamp probably washed away these metals from the sediment.

Laxmanan and Nambisan (1986) found that in mussel, *Perna viridis* gill was the major site of accumulation of mercury, copper, cadmium and lead. The accumulation of these metals in the mussel was relatively low. Ayyadurai and Krishnaswamy (1989) while studying total mercury concentration in several fresh water fishes from Madras found that the chetput swamp has slight increase in mercury with increasing depth and the concentration in water from April to July (summer) is higher which is attributed to the evaporation of water during this period. Mercury methylation from sediment to water due to bacterial activities results in higher mercury levels in bivalve molluscs (Forstner and Wittman, 1981). It is well known that mercury has high affinity for –SH groups particularly all mercury is absorbed in the form of Hg++ bond but it is likely that some may be converted by the organisms or mucosal reactions and absorbed by the animal as organic mercury. The ecotoxicological properties of metals depend principally on its capacity to form complex with free radicals with macromolecules (Boudue *et al.*, 1983). Metal residues (Hg, cd, and Pb) in benthic organisms including bivalve molluscs, though primarily depend on sediment concentrations are also influenced by complexations with particulate organic matter, which reduces the availability of heavy metals (Longston, 1986). According to Jernelov (1969) an equilibrium is existing between sediment and water with a concentration of 90–99 per cent mercury, cadmium and lead located in the sediment, while, rest is being associated with water. Paul and Pillai (1978) and Cosson (2000) have also reported that mercury and cadmium are preferentially concentrated in sediments. The higher concentrations of mercury is associated with mercury humic acid complex (Pens and Wang, 1982). In summer the mercury, cadmium and lead absorbed by humic acids in the sediments of Nandrabad swamp is probably released into the water phase (De Gregori, *et al.*, 1994). Inspite of this release metals concentration in sediment probably does not decrease because of its high retentive power (Thanabalasingam and Pickering, 1985; Ayyadurai and Krishnasamy, 1989).

Nair and Nair (1986) found seasonality in the distribution of cadmium, copper, iron, manganese, zinc, lead and mercury and that heavy metal load in the tissue was possibly due to variations in the ambient water temperature. Filter feeding bivalve molluscs take up metals rapidly from solutions or from food. Active feeding and mobilization of nutrients or energy dependent activity probably results in high accumulation of mercury, cadmium and lead in the muscular tissues like foot and adductor muscles of *L. corrianus* in monsoon and winter. Overeating of contaminated plankton and high metabolic rate resulted in higher accumulation of metals in fishes (Kurieishy, *et al.*, 1983). Skul'skii *et al.* (1987) and Kuvan (1991) found that cadmium accumulation in the mussels takes intensification of vital activity. The author further stated that significant cadmium excretion takes place not during spawning but after a certain delay at the stage of gonad reduction.

The changes in the physicochemical status of an organism when controlled by the season will superimpose itself on the trace metal availability in tile whole tissues or the defined tissues. One of the major physiological changes, which exerts a significant effect on temporal trend in the trace metal levels of higher invertebrates, is reproductive cycle (Chelomin *et al.*, 1995). These changes are directly connected with seasons, which in turn influence maturation of gametes and spawning. These physiological changes involve fluctuations in all biochemical components as well as animal weight, water contents and conditions. The loss of gametes and the loss of body weights in bivalves at spawning have consequence for trace metal seasonality (Chelomin, 1995). In this context, it is significant to note that the breeding season of *L. corrianus* in Nandrabad swamp is during monsoon and winter (from July to March). The quantity of mercury, cadmium and lead in the gonad is relatively high compared with other soft tissues in monsoon and also in winter. On the other hand, mercury and cadmium content in hepatopancreas was comparatively low in monsoon and winter which is probably related to active feeding and mobilization of metabolites to other body parts at the time the swamp gets flooded (Patil, 1993). The mantle tissue appeared to be the target organ to store more mercury, cadmium and lead in summer, monsoon and winter probably due to its direct contact will water and more surface area exposure to environment. On the other hand, in monsoon the gonad was found to contain high mercury content, which was likely due to the accumulation of organo-mercury compound with greater affinity for lipid.

From the above study it is found that the base level of mercury, cadmium and lead was high in water in monsoon and summer and in sediments lead levels was found to be higher in all the seasons. Amongst all the metals studied (Hg, Cd and Pb) mercury is proved to be most toxic to the bivalve molluscs, *Lamellidens corrianus*. The entire study reveals that the base levels of mercury, cadmium and lead in the swamp water, sediments and in bivalves, *L. corrianus* are super imposed by different seasons due to changes in environmental parameters and the changes in physiological activities of the bivalves despite the multiple human and domestic animals activities as well as domestic sewage flow and increased industrial and agricultural practices. The study helps in understanding the amount of metals being received in swamp and its biological magnification in animals residing in the swamp, particularly those at the lower level of the food chain, like *L. corrianus*. This species is consumed by poor human population as a source of protein (Patil, 1993), therefore, this study helps to make aware the local peoples for proper management of waste disposal and also to minimize biological magnification of these metals in the food chain.

References

APHA, AWWA, WPAF (1985). *Standard Methods for Examination of Water and Wastewater*, 16th Edition, APHA Publication, Washington, USA, p. 1–1267.

Ayyadurai, K. and V. Krishnaswamy (1989). Total mercury concentration in freshwater, sediment and fish. *J. Environ. Biol.*, 10(2): 165–171.

Boudou, A., Georgescauld, D. and J.P. Desmezes (1983). Ecotoxicological role of membrane barriers in transport and bioaccumulation of mercury compounds. In: *Aquatic Toxicology*, (Ed.) J.O. Nriague. John Willey and Sons, New York, p. 113–136.

Busch, D., Lucker, F. and W. Wosnjok (1998). Effect of changing salt concentrations and other physical, chemical parameters on bioavailability and bioaccumulation of heavy metals in exposed Dreissena Polymorpha. *Limnologica*, 28(3): 263–274.

Chelomin, V.P., Bobkova, E.A., Lukyanova, O.N. and N.M. Chekmasova (1995). Cadmium induced alterations in essential trace element homeostasis in the tissues of Scallop *Mizuhopecten yessoenes. Comp. Biochem. Physiol. C. Pharmacol.*

Cosson, R.P. (2000). Bivalve metallothionein as a biomarker of aquatic ecosystems pollution by trace metals: Limits and perspectives. *Cell. Mol. Biol.*, 46(2): 295–309.

Dehadrai, P.V. (1991). Pollution and aquaculture. *Nat. Symp. Env. Poll. Res. Land. and Water*, p. 17–30.

De Gregeri, I., Delgado, D., Pinochet, H., Gras N., Munoz, L., Brunn, C. and G. Navarrete (1994). Cadmium, lead, copper and mercury levels in fresh and canned bivalve mussels *Tagelus dombeii* and *Semelle Solida* from the Chilean coast.

Elder, J.F. and J.J. Collins (1991). Fresh water molluscs as indicators of bioavailability and toxicity of metals in surface-water systems. *Rev. Environ. Contam. Toxicol.*, 122: 37–79.

Frostner, U. and G.T.W. Wittmann (1981). *Metal Pollution in the Aquatic Environment*, 2nd revised edn. Springer-verlag, Berling Heidlberg, New York, p. 1–486.

Gagnon, C. and N.S. Fisher (1997). Bioavailability of sediment bound methyl and inorganic mercury to a marine bivalve. *Env. Sci. and Tech.*, 31(4): 993–998.

Hatch, W.R and L.W. Ott (1968). Determination of sub microorganism quantities of mercury by atomic absorption spectrophotometry. *J. Chemistry*, 40(14): 2087.

Jernelov. A. (1969). Mercury turnover in nature and influences on it. *Hord. Hyg. Tidskr.*, 50: 174–179 (*Chem. Abstr.*, 71: 105003).

Kureishy, T.W., Sanżgiry, S., George, M.D. and A. Braganca (1983). Mercury, cadmium and lead in different tissues of fishes and in zooplankton from the Andaman sea. *Inland J. Mar. Sci.*, pp. 12.

Kuvan, V.A. (1991). Cadmium excretion from bivalve molluscs. Dokl. Acad. NAVK. SSSR, 316(4): 1004–1006.

Lakshmanan, D.N. and Nambisan Krishnan (1986). Bioaccumulation and depuration of some trace metals by the mussel, *Perna viridis*. School of Marine Sciences, Cochin, India, p. 35–40.

Longston, W.J. (1986). Metals in sediments and benthic organisms in the mersy estuary, England, UK. *Estuarine Coastal Shelf Sci.*, 23(2): 239–262.

Lutan, J.B., Bouquet, M.M., Rauchbaar, A.B. and J. Rus (1986). Trace metal in mussel *Mytilus edulis* from coastal north sea and estuaries of EMS wester in and eastern scheldt, Netherlands. *Bull. Environ. Contam. Toxicol.*, 36(5): 770–777.

Nair, Unnikrishnan and Nair Balakrishnan (1986). Seasonality of trace metals in crassostera madrasenis inhabiting Cochin backwater. *National Seminar on Mussel Watch*, School of Marine Sciences, Cochin, India, p. 6–13.

Patil, S.S. (1993). Effect of toxic elements on the bivalves shellfishes from Maharashtra state. *Ph.D. Thesis*, Marathwada University, Aurangabad, M.S., India, p. 1–397.

Patil, S.S. and U.H. Mane (1998). Tissue biochemical levels of *Lamellidens marginalis* exposed to mercury in summer season. *J. Aqua Biol.*, 13(1&2): 73–77.

Patil, S.S. and U.H. Mane (2000). Changes in the hepatopancreas of the bivalves, *Lamellidens marginalis* due to mercury stress. *Proc. Nat. Acad. Sci. India*, 70(B), 2nd: 1–6.

Paul, A.C. and K.C. Pillai (1978). Pollution profile of river water. *Air and Soil Poll.*, 10: 133–146.

Peng, A and W. Wang (1982). Humic substances and their complex compound in natural waters. II. Complexation of humic acid with mercury in Jiyum river. *Huaning Kexve Xyebao*, 2: 214–220 (*Chem. Abstr.*, 97: 187–843).

Pringle, B.H., Hissong, D.D., Katz E.L. and S.T. Mulawka (1986). Trace metal accumulation of estuarine molluscs. *J. Saint Eng. Div. AM. Soc. CIV. Engs.*, 94: 445–475.

Rao, K.S. and M. Balaji (1991). Effect of copper on oxygen consumption on the marine fouling dressinio bivalve, *Mytilopsis sallei. Indian J. Comp. Anim. Physiol.*, 9(1)(Eng): 19–22.

Riley, J.P. and I. Roth (1974). The distribution of trace elements in some species of phytoplankton growth in culture. *J. Mar. Biol. Ass.*, UK, 51: 63– 72.

Romerill, M.G. (1971). The uptake and distribution of Zn in Oysters. *Mar. Biol.*, 9, 3: 347–354.

Singh, B.R. and E. Steinnes (1994). Soil and water contamination by heavy metals. In: *Soil Processes and Water Quality*, (Eds.) R. Lal and B.A Stewart. Lewis, Boca Raton, Florida, p. 233–272.

Sivaramakrishna, B. and K. Radhakrishnaiah (1995). Mercury accumulation in hepatopancreas and foot of *pila globosa* (snail) and *Lamellidens marginalis* at leathal and sublethal concentrations. *Proc. Ind. Nat. Sci. Acad.*, 61(6): 437–440.

Skul'skii, I.A., Pivovarova, N.B., and L.G. Kulebakina (1987). Cadmium accumulation in tissues of the mussel *Mytilus galloprovinvialis. Zh. Evol. Biokhirn. Fiziol.*, 23(3): 281–286.

Thanabalasingam, P. and W.F. Pickering (1985). The sorption of mercury (II) humic acids. *Environ. Poll. Ser. B.*, p. 267–279.

Wilcock, R.J., Smith, J.J., Pridmore, R.D., Thrush, S.F., Cummings, V.J., and J.E. Hewitt (1993). Bioaccumulation and elimination of chlordane by selected intertidal benthic fauna. *Env. Toxicol. and Chem.*, 12(4): 733–742.

Chapter 6

A Review of the Heavy Metal Toxicity on the Freshwater Fishes

M. Aruchami, A.A. Sivakumar and R. Thangam

ABSTRACT

The impact of the heavy metals on aquatic organisms is due to the movement of pollutants from various diffuse or point sources which gives rise to coincidental mixtures in the ecosystem (Anderson and D'Apollonia, 1978), thus, posing a great threat to aquatic fauna especially to fishes, which constitute one of the major sources of protein rich food for mankind. Acute and sub lethal toxic effects of heavy metals have been investigated on the fishes and their impact on aquatic ecosystem have been studied by several authors.

Keywords: *Review, Heavy metal toxicity, Freshwater fishes.*

Introduction

Aquatic environment gets contaminated with a variety of pollutants generated from diverse sources (industries, agricultural and domestic). Amongst the pollutants, pesticides, heavy metals and detergents are the major cause of concern for aquatic environment because of their toxicity, persistency and tendency to accumulate in organisms.

Heavy metals are a group of 19 elements, which have many similar physical and chemical properties and are remarkably varying from the remaining 97 known elements. Among the 19 heavy metals Lead, Cadmium and Mercury do not have any biological significance or beneficial use and are known to be extremely toxic. Other toxic metals are Chromium, Copper, Manganese, Nickel, Tin and Zinc. Once dispersed in the biosphere, these metals cannot be recovered or degraded. Hence environmental effects of metal pollution are said to be permanent.

Toxicity of Heavy Metals

Mercury

Banerjee (1985) studied the effects of heavy metal poisoning on peripheral Haemogram in *Heteropneustes fossilis* (Bloch); due to exposure to Mercuric sulphate (24-h LC_{50}). The length/breadth ratio of erythrocyte and erythrocyte nucleus did not a depart significantly from normal values and there was no significant change in erythrocyte and nuclear surface area. There was insignificant decrease in Total Erythrocyte count and Haemoglobin content but significant decrease in packed cell volume and mean cell volume. ESR showed significant increase along with MCHC but the increase in MCH was insignificant. TLC increased significantly along with large lymphocytes and basophils but small lymphocytes decreased significantly. THC increased insignificantly but CT decreased significantly.

Palanichamy and Bhaskaran (1995) studied physiological responses of the fish, *Channa straitus*. It was observed that feeding, growth, protein conversion efficiency, oxygen consumption, protein, glycogen and SDH activity of heavy metals treated *Channa striatus* decreased with increasing concentrations of Mercury, whereas the protease activity increased.

Dhande and Patil (1998) reported the effect of heavy metal pollutants on the endocrine kidney of the fish, *Channa punctatus* (Bloch). Head kidney showed increased number of inter-renal cells upto 10 days of exposure, but later on the number appeared to be slightly decreased. In Mercuric chloride treated fish the internal cells exhibited shrinkage. Chromaffin cells were distorted and lost their identify after 30 days of exposure to Mercuric chloride. Their number was increased. The hyperplasia observed in the experimental fish indicated more release of cortisol and catecholamines during the heavy metal stress to cope up with the toxic environment.

Patil and Dhande (2001) studied the effect of Mercury on ovarian recrudescence in the fresh water fish, *Channa punctatus* (Bloch). The exposure resulted in a statistically significant reduction in gonoado-somatic (GST) index after 30 days of exposure. Ovary exhibited significant decrease in cholesterol suggesting the possibility that the heavy metal either block hormone actions of the complex enzyme systems responsible for vitellogenesis.

Cadmium

Vijayram *et al.*(1989) studied the Cadmium induced changes in the biochemistry of an air breathing fish, *Anabas testudineus.* On short term exposure (3 days) a decrease was observed in liver and muscle glycogen content, accompanied by an increase in the lactic acid levels. Reduction in protein content and liver weight too was recorded. During long term exposure (54 Days) a reverse trend was observed.

Ruparalia *et al.*(1992) studied the Cadmium accumulation and biochemical alterations in the liver of freshwater fish, *Sarotherodon mossmbica* (Peters). Level of Cadmium in the liver tissue was maximum after 3 weeks exposure. Marked inhibition was noticed in the acidity of acid phosphatase during the course of study. Highest Cadmium concentration produced marked depletion of glycogen in liver at varying exposure periods. A reverse relationship between alkaline phosphatase activity and glycogen content was observed.

Cyril Arun Kumar *et al.*(1994) reported the uptake and persistence of Cadmium in tissues of the freshwater fish, *Cyprinus carpio.* Accumulation of the xenobiotic Cadmium exhibited positive correlation with concentration as well as exposure period. Persistence of Cadmium in tissues was suggested to depend on availability and persistent of the contaminant in the medium.

Palanichamy and Baskaran (1995) studied physiological responses of the fish, *Channa striatus.* It was observed that feeding, growth, protein conversion efficiency, oxygen consumption, protein, glycogen and SDH activating decreased with increasing concentrations of Cadmium whereas the protease activity increased.

Thiruvallvan *et al.*(1997) observed residual accumulation of Cadmium in gill, liver and kidney tissues of carp fingerlings of *Cyprinus carpio.* Maximum accumulation of Cadmium was observed in liver and kidney tissues respectively. Gills showed least accumulation for toxicant. The rate of accumulation was positively related with concentration of the toxicant and the period of exposure.

Sastry *et al.*(1997) observed the chronic toxic effects of Cadmium on some enzymological and biochemical parameters in *Channa punctatus.* The activities of LDH, MDH, GDH, XO, OPT and GOT were increased but the activities of SDH and PDH was reduced in gills after 15 days of exposure while the activities of all these enzymes were elevated after 30 days.

Vineeta Shukla and Sastry (1998) studied the biochemical, enzymological and haematological alterations produced on exposure of *Channa punctatus* to LC_{50} (11-2 mg/l) for 96 hr of Cadmium (Cd^{2+}). The fish were hypoglycemic and hypolactemic. The pyruvate content of blood and liver decreased in acute exposure. Depletion was noted in the total protein and glycogen content of liver and muscle, and the level of lactate in liver, while the level of muscle lactate and pyruvate increased after 96 hr. Activities of hexokinase, glucose-6-phosphatase and lactate dehydrogenase decreased in liver. The activities of hexokinase and lactate dehydrogenase in muscle and glutamate-oxaloacetate transaminase and glutamate-pyruvate transaminase in serum, liver and muscles increased. The activities of Succinate and Malate dehydrogeneases in liver and muscles decreased after acute exposure.

Rao and Manjula Sree Patnaik (1999) studied the accumulation of Cadmium in the catfish, *Mystus vittatus* (Bloch). The concentration pattern in the different organs for each metal revealed that metal in muscle were low while on other organs the levels varied. The concentration of heavy metal from the natural habitat was also analysed and there was some evidence that accumulation in different organs was influenced by the variations in pH and it was more in acidic pH.

Roshan Ara Ashram *et al.*(2003) studied the toxicity effects of Cadmium chloride on some biochemical parameters of *Symbaranchus bengalensis* and *Anabas testudineus.* Following Cadmium chloride treatment (20 ppm) the plasma protein level decreased in both the fishes. When fishes were treated with 30 ppm concentration of Cadmium chloride there was decrease in the plasma cholesterol levels of both the fishes.

Lead

Palanichamy and Baskaran (1995) studied physiological responses of the fish, *Channa striatus*. It was observed that feeding, growth, protein conversion efficiency, oxygen consumption, protein, glycogen and SDH activity were decreased with increasing concentrations of Lead, whereas the protease activity increased.

Mary Chandravathy and Reddy (1996) studied the Lead nitrate exposure changes in carbohydrate metabolism of freshwater fish, *Anabas scandens*. The changes in the activity patterns of the glycogen phosphorylases were clearly reflected in the metabolite levels. The enzymes and metabolites studied have failed to recover during initial recovery period. However on 15th day near normal levels were observed.

Rao and Manjula Sree Patnaik (1999) studied the Lead accumulation in the catfish, *Mystus vittatus* (Bloch). The concentration pattern in the different organs for Lead revealed that metal in muscle were low while on other organs the levels varied. The concentration of heavy metal from the natural habitat was also analyzed and there was some evidence that accumulation in different organs influenced by the variations in pH and it was more in acidic pH.

Nussey *et al.*(2000) studied the bioaccumulation of Lead in the tissues of the moggel, *Labeo umbratus* from Witbank Dam, Mpumalanga. Bioaccumulation of Lead varied between the gills, liver, muscle and skin. The gills generally had the highest metal concentrations, due to their intimate contact with the environment and their importance as an effector of ionic and osmotic regulation. The liver, in its role as a storage and detoxification organ can also accumulate high levels of metals. Muscles and skin accumulated much less metal concentrations. These two organs must be included in biomonitoring programmes because they are consumed by the general public. Accumulation of the metals decreased with an increase in fish length. Therefore, the smaller the fish the higher the body load of metals, due to various bioaccumulation processes.

Copper

Geetha *et al.*(1990) reported the toxic effects of heavy metal Copper on the fish, *Lepidocephalichthys thermalis*. In short term exposures more Copper in the tissue with the increased in concentration and exposure time was observed. Protein content of the fish could be taken as an indicator of Copper level.

Sastry *et al.*(1997) studied the chronic toxic effects of Copper on some enzymological and biochemical parameters in *Channa punctatus.* After 15 and 30 days in liver the activities of LDH, MDH, GDH, XO, OPT and GOT were increased but the activities of SDH and PDH was decreased. In muscle and kidney, the activities of all the enzymes were increased after 15 and 30 days of exposure to Copper. Inhibition in the activities of SDH, PDH and GDH was recorded in gills after 15 days of exposure while the activities of all these enzymes were elevated after 30 days.

Dhande and Patil (1998) studied the effect of heavy metal pollutants on the endocrine kidney of the fish, *Channa punctatus* (Bloch). Head kidney showed increased number of inter-renal cells up to 10 days of exposure, but later on the number appeared to be slightly decreased. In $CuCl_2$ treated fish the inter-renal cells exhibited shrinkage. Chromaffin cells were distorted and lost their identity after 30 days of exposure to $CuCl_2$. Their number was increased. The hyperplasia observed in the experimental fish, indicated more release of cortisol and catecholamine during the heavy metal stress to cope up with the toxic environment.

Patil and Dhande (2001) studied the effect of Copper on ovarion recrudescence in the fresh water fish, *Channa punctatus* (Bloch). The exposure resulted in statiscally significant reduction in gonado-somatic index (GSI) differ 30 days of exposure. Extensive vacuolation in the oocartex, necrosis of oolemma and hypertrophy of follicular cells, degeneration of vitellogenic and hypertropy of follicular cells, degeneration of vitellogenic and large vitellogenic oocytes were observed. Further in fish the ovary exhibited significant decrease in cholesterol suggesting possibility that the heavy metal either block hormone actions or the complex enzyme systems responsible for vitellognensis.

Chromium

Sankar and Konar (1995) studied the effects of Chromium on fish and fish food organisms. It was observed that fish at sublethal concentration were reduced significantly; growth and reproduction of fish were hampered by sublethal level.

Nussey *et al.* (2000) studied the bioaccumulation of Chromium in the tissues of the moggel, *Labeo umbratus* from Witbank Dam, Mpumalanga. Bioaccumulation of Chromium varied between the gills, liver, muscle and skin. The gills generally had the highest metal concentrations, due to their intimate contact with the environment and their importance as an effecter of ionic and osmotic regulation. The liver, in its role as a storage and detoxification organ, can also accumulate high levels of metals. Muscles and skin accumulated much less metal concentrations. Accumulation of the metal decreased with an increase in fish length.

Manganese

Nussey *et al.* (2000) studied the accumulation of Manganese in the tissues of the moggel, *Labeo umbratus* (*Cyprinidae*) from Witbank Dam, Mpumalanga. Bioaccumulation of Manganese decreased with increase in length of the fish.

Nickel

Nussey *et al.* (2000) studied the bioaccumulation of Nickel in the tissues of the moggel, *Labeo umbratus* (*Cyprinidae*), from Witbank dam, Mpumalanga. Bioaccumulation of Nickel decreased with increase in length of the fish.

Iron

Arun *et al.* (2000) observed the reduction in biochemical parameters (protein, carbohydrate, lipid) of the fish, *Oreochromis mossambicus* when exposed to Iron for 15 days. The detoxification mechanism in *Oreochromis mossambicus* was scrutinised by monitoring the induction of xenobiotic conjugating enzyme (Glutathione-s-transferase) at every 5 days interval.

Zinc

Arun *et al.* (2000) observed the reduction in protein, carbohydrate and lipid during exposure of Zinc *Oreochromis mossambicus.* Mori *et al.* (1990) studied the influence of water hardness and heavy metal (Zinc) on food utilization in *Cyprinus carpio* var. communis. It was noted that consumption, absorption, conversion and metabolism of *Cyprinus carpio* reared at 0.4 mg concentration of Zinc with 160 and 768 mg $CaCO_3$/l decreased, whereas at 384 and 576 $CaCO_3$/l were increased. The efficiency of absorption and conversion also followed the same trend.

Rao and Manjula Sree Patnaik (1999) studied the concentration pattern in the different organs for Zinc revealed that metal in muscle were low while on other organs the levels varied.

Discussion

Fishes are exposed to elevated metal levels in an aquatic environment; they can absorb the bioavailable metals directly from the environment via, the gills and skin or through the injection of contaminated water and food. Metals in the fish are then transported by the bloodstream, which brings it into contact with the various organs and tissues (Vander Putte and Part, 1982). Fish can regulate metal concentration to a certain extent; whereafter bioaccumulation will occur (Health, 1991). Therefore, the ability of each tissue to either regulate or accumulate metals can be directly related to the total amount accumulated in that specific tissue. Furthermore, physiological differences and the position of each tissue in the fish can also influence the bioaccumulation of a particular metal (Kortze, 1997).

Mercury has insignificant decrease in total erythrocyte count and Haemoglobin content but significant decrease in packed cell volume and mean cell volume (Banerjee, 1985). The increase in the concentration of Mercury leads to decrease in the oxygen consumption, protein, glycogen and SDH activity, excepting the protease which has increased activity (Palanichamy and Bhaskaran, 1995). Mercuric chloride affects the internal cells of the kidneys of the fishes. Initially the cells shrink and after continuous exposure chromaffin cells can be destorted as observed by Dhande and Patil (1998). The gonado-somatic index and the cholesterol content in the ovaries are also decreased.

Cadmium decreases the glycogen content in the muscle and liver and protein in the muscle (Vijaram *et al.,* 1989). Cadmium inhibits the acid phosphatase activity when present in the liver (Ruparalia *et al.,* 1992). Cadmium behaves similar to the Mercury when present in the tissues of the fishes by showing decrease in feeding, growth, oxygen consumption, protein, glycogen and SDH activity. Cadmium increases LDH, MDH, GDH, XO, GPT and GOT at the same time decreases the activities of SDH and PDH (Sastry *et al.,* 1997).

Lead behaves similar to that of Mercury and Cadmium by showing reduction in the feeding, growth, protein conversion efficiency, excepting the protease activity which increases (Palanichamy and Bhaskaran, 1995). Gills of fishes accumulate more amount of Lead than liver, muscle and skin, due to their intimate contact with the environment and their importance as an effector of ionic and osmotic regulation. The liver can also accumulate more levels of heavy metals as a storage and detoxification organ (Nussey *el al.*, 2000).

Copper increases the activities of LDH, MDH, GDH, OX, GPT and GDT whereas decreases the SDH and PDH activities as observed by Sastry *et al.* (1997). Copper induces hyperplasia in the kidney cells of the fishes and releases more of cortisol and catecholamine due to this metal stress (Dhande and Patil, 1998).

Chromium affects the growth and reproduction of the fishes even in the sub-lethal level as observed by Sankar and Konar (1995).

Manganese and Nickel behave similar to the Chromium when present in the tissues of fishes. Higher amounts of Manganese and Nickel are found to be accumulated in gills and liver (Nussey *et al.*, 2000).

Iron and Zinc decrease the protein, carbohydrate and lipid content of the muscle and gill of fishes (Arun *et al.*, 2000).

There are clear indications that heavy metals in the environment have increased to harmful levels. Exposure to these heavy metals show adverse biological conditions. Occupational exposure to heavy metals causes dermatitis, perforation of the nasal septum, and inflammation of the larynx and live in human beings. Hence biomonitoring is necessary for safe environment of the future.

References

Arun, S., Thirumurugan, R., Visakan, R., Balamurugan, S., Arivazhagan, R., Subramanian, P., Prince Vijeya Singh and T. Malarvizhi (2000). Toxicity induced biochemical modulations and phase II xenobiotic conjugating enzyme (GST) in *Oreochromis mossambicus. Asian. J. S. Microbial. Biotech, and Environ. Sci.*, 2(34): 326–330.

Ashram, Roshan Ara, Manjula Yadav, Dillep Kumar and B.N. Yadav (2003). Toxicity effects of Cadmium chloride on some biochemical parameters of *Symbaranchus bengalensis* and *Anabas testudineus*. *Bio-Science Res. Bull.*, 19(1): 75–78.

Banerjee, V. (1985). Effect of heavy metal poisoning on peripheral haemogram in *Heteropneustes fossilis* (Bloch): I. Mercuric sulphate (LC_{50}). *Comp. Physiol. Ecol.*, 11(4): 173–176.

Dhande, R.R. and G.P. Patil (1998). Effect of heavy metal pollutants on the endocrine kidney of the fish, *Channa punctatus* (Bloch). *Proc. Acad. Environ. Biol.*, 7(2): 149–154.

Geetha, R., Kumaraguru, A.K. and A.J. Thatheyus (1996). Toxic effects of heavy metal Copper on the fish, *Lepidocephalichthys thermalis* in short term exposures. *Poll. Res.*, 15(20): 151–153.

Health, A.G. (1991). *Water Pollution and Fish Physiology*. Lewis Publishers, Boca Raton, Florida, USA, pp. 359.

Kortze, P.I. (1997). Aspects of water quality, metal contamination of sediment and fish in the Olifants River, Mpumalanga. *M.Sc. Thesis*, Rand Afr. Univ. South Africa.

Kumar, Cyril Arun L., Vincent, S. and T. Ambrose (1994). Uptake and persistence of the heavy metal Cadmium in tissues of the freshwater fish, *Cyprinus carpio*. *Poll. Res.*, 13(4): 361–364.

Mary Chandravathy, V. and S.L.N. Reddy (1996). Lead nitrate exposure changes in carbohydrate metabolism of freshwater fish. *J. Environ. Biol.*, 7: 75–79.

Moni, D., Baskaran, P. and S. Palanichamy (1990). Influence of water hardness and heavy metal (Zinc) on food utilization in *Cyprinus carpio*. Var. communis. *J. Ecobiol.*, 2(1): 51–55.

Nussey, G., J.H.J. Van Vuren and Preez (2000). Bioaccumulation of Chromium, Manganese, Nickel and Lead in the tissues of the moggel, *Lebeo umbratus* (Cyprinidae), from Witbank Dam, Mpumalanga. *Water SA*, 26(2): 269–289.

Palanichamy, S. and P. Baskaran (1995). Selected biochemical and physiological responses of the fish, *Channa punctatus* as biomonitor to assess heavy metal pollution in freshwater environment. *J. Ecotoxicol. Environ. Monit.*, 5(2): 131–138.

Patil, G.P. and R.R. Dhande (2001). Effect of Mercury and Copper on ovarian recrudescence in the freshwater fish, *Channa punctatus* (Bloch). *J. Ecotoxicol. Environ. Monit.*, 11(2): 83–90.

Rao, L.M. and R. Manjula Sree Patnaik (1999). Heavy metal accumulation in the catfish, *Mystus vittatus* (Bloch) from Mehadrigedda stream of Visakhapatnam, India. *Poll. Res.*, 19(3): 325–329.

Ruparelia, S.G., Yogendra Verma, Meta, N.S. and U.M. Rawal (1992). Cadmium accumulation and biochemical alternations in the liver of freshwater fish, *Saratherodon mossambica* (Peters). *J. Ecotoxicol. Environ. Monit.*, 2(2): 129–136.

Sarker, U.K. and S.K. Konar (1995). The combined effects of pesticide Thiodan, heavy metal Chromium and detergent Ekaline FI on fish and fish organisms. *J. Environ. Biol.*, 19(1): 19–26.

Sastry, K.V., Sarita Sachdeva and Pratima Rathee (1997). Chronic toxic effects Cadmium and Copper, and their combination on some enzymological and biochemical parameters in *Channa punctatus. J. Environ. Biol.* 18(3): 291–303.

Shukla, Vineeta and K.V. Sastry (1998). Acute toxic effects of Cadmium on various parameters in the fresh water fish. *Proc. Acad. Environ. Biol.*, 7(2): 155–160.

Thiruvalluvan, M., Nagendran, N. and A. Charles Manoharan (1997). Bioaccumulation of Cadmium and Methylparathion in *Cyprinus carpio* Var. communis (Linn.). *J. Env. & Poll.*, 4(3): 221–224.

Vander Putte, I. and P. Part (1982). Oxygen and Chromium transfer in perfused gills of rainbow trout, *Salmo gairdneri* exposed to hexavalent Chromium at two different pH levels. *Aquatic. Toxicol.*, 2: 31–45.

Vijayram, K., Geraldine, P., Varadarajan, T.S., George John and P. Loganathan (1989). Cadmium induced changes in the biochemistry of an air breathing fish *Anabas testudineus. J. Ecobiol.*, 1(4): 245–251.

Chapter 7

Occurrence of Heavy Metals in Tapi River Water and Sediments at Bhusawal Region

R.B. Dhake and S.T. Ingle

ABSTRACT

The concentration of Fe, Mn, Zn, Cu, Pb, Ni and Cd were determined in water and sediments of the river Tapi at Bhusawal region. The analytical results shows that relatively high concentrations of these metals in water and sediment samples were obtained at Bhusawal where huge quantity of domestic and industrial waste effluent discharged into the river. The trace metal content in sediments were found higher as compared with water. The iron is found in extremely high amount and Cadmium content was lowest than other heavy metals.

Keywords: *Tapi river, Water pollution, Heavy metal, Sediment.*

Introduction

River Tapi is the most important river in North Maharashtra. It originates from Baitul (M.P.), is a west flowing river covering approximately 247 K.M. in Maharashtra state. The main natural source of metals in the river water are weathering of minerals. Industrial effluents, non-point pollution sources, as well as atmospheric precipitation (Solomons and Forstner, 1984) can also be sources of increased concentration of heavy metals. The contamination of river can be caused due to the untreated or inadequately treated municipal and industrial effluents and other human activities like bathing, washing of clothes and immersing of ashes. Trace elements, especially so-called heavy metals are among the most common environmental pollutants. Their occurrence in water and biota indicate the presence of natural or anthropogenic sources (Forstner and Wittman, 1979). In cases of anthropogenic pollution, substantial metal accumulation in sediments may take place.

The pollutants, mainly heavy metals and some hydrocarbons discharged from industries and sewage into the river normally come with water in the form of colloids, suspensions, and particulates. In due course of time the pollutants deposit in river beds and get mixed with other water-borne sediments. The deposition of pollutants/ heavy metals is size dependant, generally occurs to be associated with very fine particles of silt and clay. The mechanism of adsorption and deposition of heavy metals would take place due to several ongoing physico-chemical reactions in water system. Heavy metal concentrations in water and aquatic eco-systems have been analysed worldwide (Borg and Johansson, 1989; Runnels *et al.*, 1992 Gadh *et al.*, 1993), but there are few publications on their concentrations in Tapi river.

The occurrence of elevated levels of trace metals, especially in the sediments can be good indication of man induced pollutions and high levels of heavy metals can often be attributed to anthropogenic influences rather than natural enrichments of the sediments by geological weathering (Davis *et al.*, 19991 and Lord and Thompson, 1988). The objectives of the present study were to determine heavy metal concentrations and their speciation forms in sediment of Tapi river.

Materials and Methods

Water and surface sediments samples were collected in Tapi river basin at Bhusawal. Figure 7.1 shows the sampling locations of the present study. Based on the human activity and industrial locations, three locations along the stretch of river were selected for the present study in the period of 2001–2002. Station–I (Mahadev hill), presents clean water, free from waste discharges and no

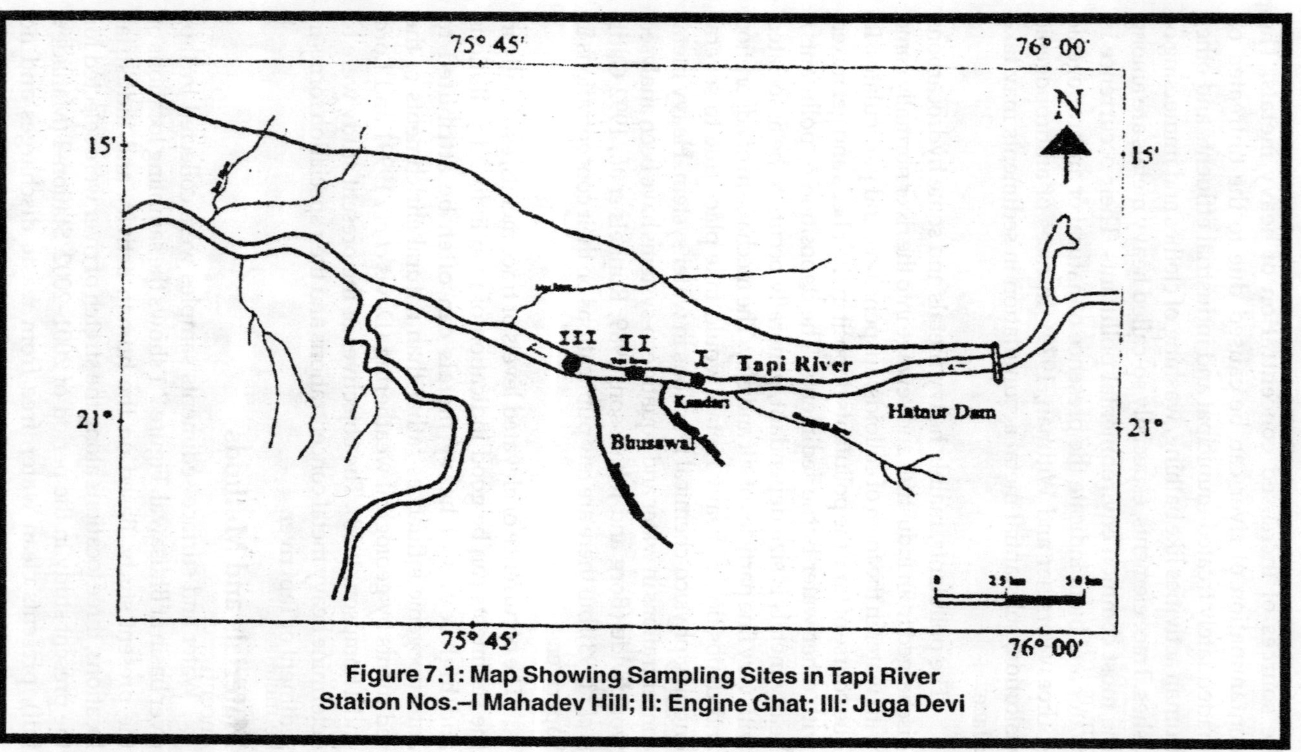

Figure 7.1: Map Showing Sampling Sites in Tapi River
Station Nos.–I Mahadev Hill; II: Engine Ghat; III: Juga Devi

industries are located near by this areas, hence considered as control for present study. Station–II (Engine Ghat), this station represents the region affected by municipal sewage discharge from Bhusawal township. From this station water is supplied to the town for human usage. Station–III (Juga Devi), this is the last station affected by municipal sewage discharge, which contains industrial wastes coming from the industries.

For water and sediments samples were collected and analysis was conducted following the standard methods of APHA (1989). For each sampling station, 100 ml of water samples were acidified with 2 ml Conc. HNO_3 and all samples were concentrated up to 10 ml using evaporation method. The sediment samples were crushed in a porcelain mortar to pass through 80 µg mesh sieve. One gram of prepared sample was digested with 25 ml Conc. HNO_3 and the residue was diluted with 10 ml deionised distilled water. All the digested samples of water and sediments were analysed for different heavy metals (Fe, Mn, Zn, Cu, Pb, Ni, and Cd) by Atomic Absorption Spectrophotometer UNICAM969.

Results

It is the general observation for all sampling stations that Fe and Mn are present in higher concentrations than other metal (Table 7.1).

Metal Analysis of Water Sample

Station–I

Compared to other stations, the river water of Station–I should show low levels of all the metals. Lead and Cadmium were found in much lower quantities. Pb ranged between 5.87–7.71 µg/lit, whereas Cd was 1.30–1.86 µg/lit. Nickel (Ni) ranged between 7.25–14.55 µg/lit. Iron (Fe) is the most abundant metal among all the metals studied and ranged between 226–550.50 µg/lit at this station. Manganese (Mn) is the second element in higher concentration and varied between 64.50–136.75 µg/lit in the water of Station–I. Zinc (Zn) ranged between 18.73–73.30 µg/lit and copper (Cu) ranged between 11.56–28.30 µg/lit.

Station–II

The higher values of Zn were recorded at Station–II as compared with other stations (I and III) Zn varied between 69.08–149.78

Table 7.1: Levels (μg/lit.) of Metals in Water Sample of the Tapi River

Metal	*Station–I*		*Station–II*		*Station–III*	
	Mean	*Range*	*Mean*	*Range*	*Mean*	*Range*
Fe	375.06 ± 141.38	226.00–550.50	556.56 ± 187.26	366.25–788.50	866.88 ± 259.26	622.50–1208.00
Mn	95.50 ± 33.05	64.50–136.75	198.50 ± 95.61	75.00–298.25	312.65 ± 119.20	147.88–404.78
Zn	46.70 ± 24.27	18.73–73.30	109.38 ± 36.93	69.08–149.78	79.74 ± 35.93	40.15–120.83
Cu	20.53 ± 7.63	11.56–28.30	43.96 ± 10.18	31.80–55.10	63.11 ± 10.52	51.65–74.15
Pb	6.78 ± 1.10	5.84–7.71	9.39 ± 7.37	4.58–20.25	11.03 ± 4.94	6.00–17.28
Ni	11.19 ± 3.67	7.25–14.55	7.82 ± 2.02	6.05–10.41	9.96 ± 4.61	5.24–15.40
Cd	1.58 ± 0.39	1.30–1.86	2.76 ± 1.06	1.56–4.02	5.25 ± 1.82	3.46–7.62
pH	8.25 ± 0.10	8.15–8.36	8.62 ± 0.06	8.55–8.70	8.45 ± 0.18	8.20–8.59

Values are ± SD of 16 determinations.

Table 7.2: Levels (µg/lit.) of Metals in Sediments of the Tapi River

Metal	*Station–I*		*Station–II*		*Station–III*	
	Mean	*Range*	*Mean*	*Range*	*Mean*	*Range*
Fe	2024.38 ± 211.15	1779.00–2259.50	2477.38 ± 128.43	2339.50–2628.25	2299.13 ± 174.96	2106.50–2497.75
Mn	71.25 ± 46.32	18.35–115.55	197.50 ± 67.93	123.75–278.25	250.00 ± 63.45	198.00–340.25
Zn	289.69 ± 81.73	196.25–381.25	415.19 ± 102.03	290.00–524.00	384.31 ± 122.06	273.50–544.00
Cu	42.13 ± 12.27	28.50–56.50	71.06 ± 17.38	51.50–90.50	84.69 ± 29.60	52.75–120.00
Pb	11.07 ± 2.23	9.09–13.73	18.60 ± 5.77	12.48–25.53	20.34 ± 7.85	12.26–30.15
Ni	1.33 ± 0.43	0.89–1.73	2.11 ± 0.62	1.45–2.80	33.79 ± 9.59	23.03–44.08
Cd	1.27 ± 1.47	0.28–2.61	0.80 ± 0.45	0.41–1.41	1.36 ± 0.82	0.66–2.45

Values are ± SD of 16 determinations.

μg/lit. Fe varied between 366.25–788.50 μg/lit whereas Mn 75.00–298.25 μg/lit in the water of Station–II. At Station–II Pb and Cd were present in low concentrations and Cu found to be between 31.80–55.10 μg/lit whereas Ni 6.05–10.41 μg/lit in the water of Station–II.

Station–III

There are remarkable changes in the metal levels of Station–III as compared to other stations (I and II) except for Zn. The highest concentrations of Fe, Mn, and Cu at Station–III were recorded which were higher as compared to other stations. Fe, Mn, and Cu ranged between 622.50–1208.00, 147.80–404.78 and 40.15–120.83 μg/lit respectively in the water of this station. Pb, Ni, and Cd levels for Station–III water samples were found to be more less same as that of Station–II.

Metal Analysis of Sediment Samples

Station–I

At Station–I, the sediment samples showed trace amount of Cadmium. It ranged between 0.28–2.61 μg/g where as Ni varied between 0.89–1.73 μg/g. Fe and Mn were higher in the sediments as compared to other stations (II and III). Fe and Mn varied between 1779–2259.59 and 18.35–115.55 μg/g in the sediments respectively. As compared to other stations, lower concentrations of Zn, Pb, and Cu were recorded at Station–I. Zn, Pb and Cu ranged between 196.25–381.25, 9.09–13.73 and 28.50–56.50 μg/g respectively in the sediments of this station.

Station–II

At Station–II, Fe has highest recorded value which is more in comparison with Stations–I and III. Fe varied between 2339.5–2628.25 μg/g. The levels of Mn, Zn and Cd ranged between 123.75–278.0, 290–524 μg/g and Cd 0.41–1.41 μg/g respectively were highest recorded values in the sediments. Higher levels of Fe, Mn, Zn, Cu, Pb, Ni and Cd were observed in the sediments of this station as compared to Station–I.

Station–III

At Station–III, Ni 44.08 μg/g has the highest value which is more in comparison with Stations–I and II values. The higher values of Mn, Zn, Cu and Pb were recorded at Station–III as compared to other stations (I and II) and ranged between 198–340.25, 273.5–544.0, 52.75–120.0 and 12.26–30.15 μg/g respectively in the sediments of this station.

Discussion

The industrial waste effluents along with sewages from the manufacturing of paints and pigments are discharged in Tapi river at sampling Stations–II and III. Ferromanganese alloys, storage batteries, ceramics and iron foundries generates lot of waste water is being added to the river which may contribute to higher concentrations of Fe, Mn, and other heavy metal in water and sediment samples.

Generally the concentrations of heavy metals in sediments were found considerably higher than those obtained from water sample. The higher concentrations of Fe, Mn and other heavy metals in sediments as compared to water sample, may be attributed to the precipitation of metals as insoluble hydroxides through oxidation and hydrolysis under alkaline pH condition (Stumm and Morgan, 1970).

The pH survey of the Tapi river water at all sampling stations reveals that the river water is alkaline. Generally the precipitation of trace metals as carbonates, oxides and hydroxides occurs at higher pH and sink down as sediments (Dean *et al.*, 1972). The concentrations of Cu, Cd, Zn and Pb metals were found to be higher in sediments than water samples at every sampling station may be due to precipitation of these metals as carbonates, (Ingle *et al.*, 1993) have also reported precipitation of Calcium carbonates along with heavy metals from solution under high pH condition. Popova (1961) demonstrated that heavy metal carbonates of low solubility such as $CaCO_3$ and $PbCO_3$ are completely eliminated from the solution as a result of precipitation.

Metals like Cu, Mn and Ni have interact with the organic matter in aqueous phase and settle down, resulting in high concentration in sediments (Elewa *et al.*, 1990). Mineralogical studies of polluted sediments indicate that heavy metals were found to be associated with finer particles of slit and clay which have sufficiently large area and size to accumulate metal ion due to their intermolecular forces known as adsorption (Baruah *et al.*, 1996).

Rapid industrial growth, urbanization and agricultural developments are the main factors responsible for the present river water pollution in the study area. Current investigation provides exclusive evidence of large scale deterioration of river water quality as measured by selected heavy metals concentrations. Efficient and appropriate monitoring efforts are needed in the area as it is the most susceptible to water contamination or degradation.

References

A.P.H.A. (1989). *Standard Method for the Examination of Water and Wastewater Including Bottom Sediments and Sludge*, 19th Ed., New York.

Baruah, N.K., Kotok, Y.P., Bhattacharya, K.G. and G.C. Boroh (1996). Metal speciation in Jhanji River. *Sediments. Sci. Total Env.*, 193(1): 1–21.

Borg, H. and K. Johansson (1989). Metal fluxes to Swedish forest lakes. *Water, Air & Soil Poll.*, 47: 427–440.

Davies, C.A., Tomlinson, K. and T. Stephenson (1991). Heavy metals in river fees estuary sediments. *Env. Tech.*, 12: 961–972.

Dean, J.G., Bosqui, F.L., and K.H. Lanouetle (1972). Removing heavy metals from wastewater. *Env. Sci. Tech.*, 6: 518–522.

Elewa, A.A., Sayyah, S.H. and A. Fouda (1990). Distribution of some pollutants in lake. Nasser and river Nile at Aswan., Reg. Smp. Environ. Study. (MNARC), Alex., p. 382–402.

Forstner, U. and G.T.W. Wittman (1979). *Metal Polluting in the Aquatic Environment*. Springer, Berlin, New York.

Gadh, R., Tandon, S.N., Mathur, R.P. and Singh (1993). Specification of metals in Yamuna river. *Sediments Sci. Total Env.*, 136: 229–242.

Ingle, S.T., Vishwarajan S., Suryawanshi, S.A. and B.G. Kulkarni (1993). Impact of industrialization of heavy metal concentrations of the Patalganga river, Maharashtra. *Env. & Ecol.*, 11(4): 747–752.

Lord, D.A. and G.A. Thompson (1998). The swartkops estuary, pollution status. In: Proc. Symp. Held on 11 and 15 September 1998 at the University of Port Elizabeth, (Eds.) Baid, D., Marais,

J.F.K. and A.P. Martin. S. Afr. Nah. and Sci. Programmes Rep. No. 156: 16–24.

Popova, T.P. (1961). Coprecipitation of some microconstituents from natural water and Calcium carbonate, *Geochemistry*, 12: 1256–1261.

Runnels, D.C., Sheperd, T.A. and E.A. Angino (1992). Metals in water determining natural back ground concentrations in mineralized areas. *Env. Sci. Tech.*, 26: 2316–2323.

Solomons, W. and U. Forstner (1984). *Metals in the Hydrocycle.* Springer, Berlin, pp. 653.

Stumm, W. and J.J. Morgan (1970). *Aquatic Chemistry*. Wiley, New York.

Chapter 8

Bioaccumulation of Mercury in Uppanar Estuary, Cuddalore, Southeast Coast of India

R. Rajaram and M. Srinivasan

ABSTRACT

Estuarine ecosystem are highly complex, dynamic and subject to many internal and external relationship that are subject to change over time. Uppanar estuary is considered to be one of the highly polluted estuary in south east coast of India due to industrialization. SIPCOT (Small Industrial Promotion Corporation of Tamil Nadu, covering an area of about 520 acres with 44 industries) is located on the bank of Uppanar estuary at Cuddalore. It was established for chemical, petrochemical, pharmaceutical, biocides, fertilizer, fungicides, chlor-alkali and metal processing industries etc. Indiscriminate discharges of waste water from SIPCOT industrial complex into coastal environment affect both biotic and abiotic system and finally cause some ill effects to human beings through food-chain. A detailed study was made on the bioaccumulation of mercury in the food-chain and the results were discussed in this article.

Keywords: *Mercury, Food-chain, Uppanar Estuary.*

Introduction

Coastal zone comprises varied biotopes such as estuaries, backwaters, mangroves, salt marshes, coral reefs, lagoons and near shores. Estuaries are highly protective, sensitive, feeding and breeding ground for fin and shell fishes with rich biodiversity. Realizing the importance of the estuaries, the ever exploding human population exploit not only the biological resources but also interferes and modifies the basic coastal processes. This results in many environmental episodes such as "Minamatta" and "Itai Itai" diseases.

The coastal zone receives the pollutants principally from three major routes–*viz.* atmosphere, riverine and glaciers. Of course man also serve as a "Geological Agent" by the way of discharging the effluents through piped outfalls, direct pimping, operation of ships etc. When the pollutants exceed the threshold limits coupled with environmental variables such as salinity, temperature, pH, DO, H_2S etc. give stress to organisms of the environment. So monitoring of estuarine environment needs be carried out periodically to estimate the concentration and distribution of varies pollutant levels to observe the changes in the ecosystem. Discharges of waste water from SIPCOT industrial complex estimated to be 56.92 gallons in lakhs/day into the coastal environment. So the present study were, therefore, initiated to understand the mercury distribution pattern in Uppanar estuarine food-chain with regard to effluents from SIPCOT industries.

Sources of Mercury

Natural Occurrences

2700–6000 tonnes/year due to earth's crust, emission from volcanoes and evaporation from natural water bodies (Food and Drug Administration).

Anthropogenic Sources

3000 tonnes are released annually into the atmosphere by human activities (Judith, 2000). About 180 tonnes of mercury and its compounds are introduced into Indian coastal environment per annum (Patel and Chandy, 1988) by the following industries:

1. Pharmaceutical and cosmetic industries
   (a) Phenyl mercury acetate used in contraceptive vaginal jellies.
   (b) Various organo-mercury compounds used as diuretics.
   (c) Organo-mercurial are used as antiseptic products.

2. Electrical apparatus industries

   Mercury was used in the manufacture of mercury batteries, mercury pool rectifiers and power tubes, varieties of lamps includes fluorescent, germicidal, photocopying and high intensity are discharge lamps.

3. Industrial control instrument industries

   It is used in the manufacture of switches, relays, gauges, pump seals and values.

4. General laboratory

   Mercury used in experimental equipments such as diffusion pumps, barameters, monometers, thermometers and vibration dampers.

5. Chlor-alkali industries uses continuous flow of mercury cathode cell to produce chlorine and caustic soda

6. Paint industries

   (a) Organo-mercurial compounds: Bactericide-fungicide agents to protect water based paints from bacterial fermentation

   (b) Phenyl mercury derivatives are used as paint preservatives

7. Amalgam alloy is made of silver, tin and copper mixed with mercury to form plastic mass used to fill the tooth cavities

8. Mercury is used to prepare various catalytic salts such as chloride, oxide, sulfate, acetate etc.

9. Various mercury compounds are used as slimicides to control microbial growth in paper and pulp industries.

Materials and Methods

Uppanar estuary (Lat. 11° 42′N; Long. 79° 46′E) is formed by the confluence of Gadilam and Paravanar river, and opens in to the Bay of Bengal near Cuddalore old town on the south east coast of India (Figure 8.1). It forms a potential fishing ground with an annual average landing of about 2000 tonnes. The tidal effect extends up to a distance of about 6 km and width of the estuary is about 30m near the mouth and 20 m in the upstream. SIPCOT industrial complex is located on the bank of Uppanar estuary, where the effluents are

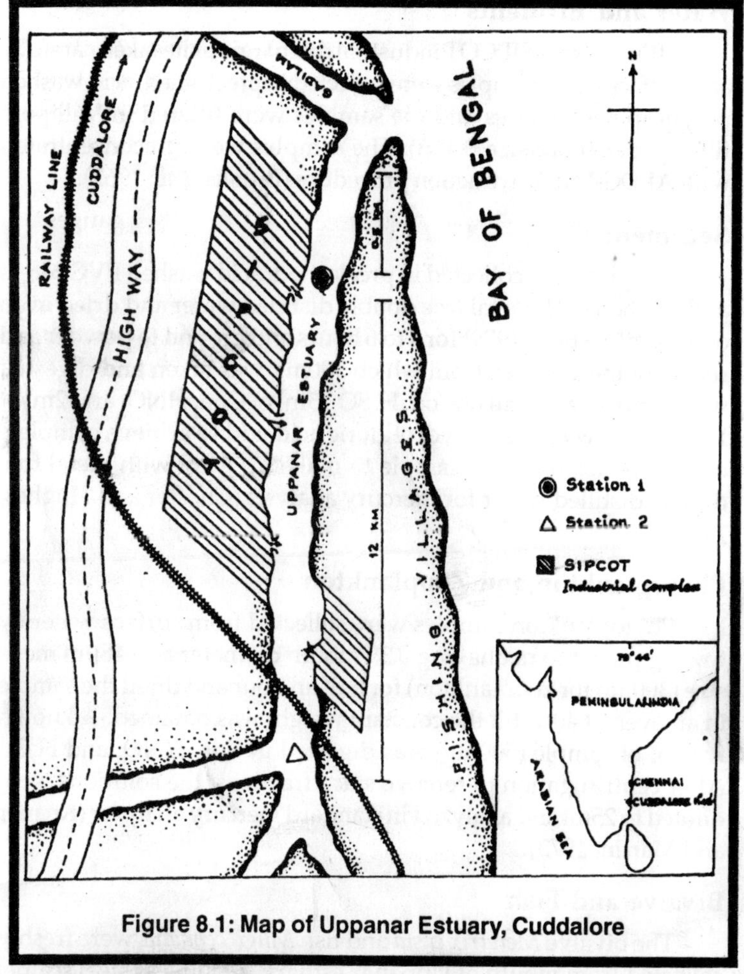

Figure 8.1: Map of Uppanar Estuary, Cuddalore

discharging, mixing and diluting. Apart from the industrial effluents, drainage of municipal and domestic sewage from Cuddalore old and new towns and waste from coconut husk retting ground are discharging. The bottom soil is characterized by blackish clay in nature, salinity ranges from 5 to 36 ppt. Sediment, phytoplankton, zooplankton, bivalve, fish, water and effluent were collected from the study areas for the analysis of mercury.

Water and Effluents

Effluent from SIPCOT industrial discharging site taken carefully and surface water samples were collected in precleaned, acid washed polypropylene bottles and the samples were filtered in Millipore filter paper (Pore size 0.45µ). The samples were preconcentrated with APDC-MIBK extraction procedure (Brooks *et al.*, 1967).

Sediment

Samples were collected in precleaned, acid washed PVC corers and washed with metal free double distilled water and dried in an oven at 40°C (EPA, 1979) for 5 to 6 hours and ground to powder and redrying the samples from which 500mg was taken and digested with a mixture of 1 ml of Con. H_2SO_4, 5ml of Con. HNO_3 and 2ml of $HClO_4$. A few drops of hydrofluoric acid add to achieve complete digestion and filter the sample to make up 25ml with metal free double distilled water for mercury analysis (Chester and Hughes, 1967).

Phytoplankton and Zooplankton

Phytoplankton samples were collected from surface water by towing plankton net having 0.35 mouth diameter and 48µm mesh size (300µm for zooplankton) for half an hour and dried the sample in an oven at 40°C till the constant weight was obtained. A known weight of sample (500mg) was digested in Con. HNO_3 and H_2O_2. After centrifugation to remove silica fractions, the solutions were diluted to 25ml and analyzed in standard mercury analyzer (Knauer and Martin, 1973).

Bivalve and Fish

The bivalve *Meretrix casta* and fish *Mugil cephalus* were freshly collected from mouth of Uppanar estuary. A stainless steel scalpel steel was used to remove foot and muscle from bivalve and gill, liver, kidney and tissue from fish and the dissected portions were dried in hot air oven at 40°C. The dried samples were powdered and digested in 3 : 1 ratio of HNO_3 and $HClO_4$. The digested samples were diluted to make up 25ml and stored for mercury analysis (Topping, 1973).

Total mercury was quantified by adopting cold vapour technique in mercury analyzer, model No. 5800D, Serial No. 209, fabricated by Electronic Corporation of India, Bangalore.

Results

Distribution and accumulation of mercury in water, effluent, sediment, phytoplankton, zooplankton and various organs in bivalve and fish are showed in Table 8.1.

Table 8.1: Bioaccumulation of Mercury in Uppanar Estuary

Sl.No.	*Samples*	*Value*
1.	Effluent	0.17–11.49 $\mu g\ l^{-1}$
2.	Water	N.D–1.92 $\mu g\ l^{-1}$
3.	Sediment	0.17–6.710 $\mu g\ g^{-1}$
4.	Phytoplankton	0.12–6.87 $\mu g\ g^{-1}$
5.	Zooplankton	0.49–8.87 $\mu g\ g^{-1}$
6.	*Meretrix casta*–Foot	N.D–6.42 $\mu g\ g^{-1}$
	–Body muscle	N.D–5.48 $\mu g\ g^{-1}$
7.	*Mugil cephalus*–Gill	0.08–3.1 $\mu g\ g^{-1}$
	–Liver	0.11–3.92 $\mu g\ g^{-1}$
	–Kidney	N.D–2.1 $\mu g\ g^{-1}$
	–Tissue	N.D–0.23 $\mu g\ g^{-1}$

N.D.: Non-detectable value.

Discussion

Natural sources of mercury in water are through land and river runoff and mechanical and chemical weathering of rocks. The high value of mercury in water due to rainfall causing flooding the stagnant effluent and facilitating higher influence of land derived materials including agricultural runoff and washing of waste coupled with river discharges and direct discharge of effluents. The estuarine sediments may be exchanging part of the exchangeable phase of the metals in water. The high value of mercury in sediment may be due to land drainage, microbial activity, sediment particle type and size and organic content which influence the absorption and accumulation of mercury in sediments.

Many environmental factors influence and affect the uptake of mercury by phytoplankton. This includes physico-chemical, biological and bio-chemical parameters and also the interaction between the mercury by environment. Phytoplankton should have concentrated more heavy metal by adsorption and absorption mechanisms from the surface waters (Knauer and Martin, 1973; Bryan, 1976). Mercury accumulation by zooplankton is mainly by two pathways *i.e.*, direct uptake from the water through and the assimilation from ingested food and detritus (Davis, 1978). High concentration of mercury in zooplankton due to land runoff, heavy fresh water flux, industrial effluents and domestic sewage (Knauer and Martin (1972) stated that zooplankton ingested food (phytoplankton) with high mercury content seems to increase the concentration of mercury in zooplankton.

Accumulation of metals by mussel is influenced by salinity, temperature and concentration of metal in water (Boyden and Romeril, 1974). The size, sex, reproductive condition and season are also factors for the variation. In the present study the accumulation of mercury in *Meretrix casta* showed N.D to 6.42 $\mu g\ g^{-1}$ and more amount of mercury accumulation due to filter feeding habit (Paul Raj, 1987).

Fishes are known to concentrate metals in their body tissues in varying proportions depending upon the species, environmental conditions and feeding habits and accumulates high values than their basal environment due to the process of bioamplification (Cooper, 1983). The bioaccumulation of mercury in fish *Mugil cephalus* showed N.D–3.92 $\mu g\ g^{-1}$ may be due to behavior of the animal with respect to their environment *i.e.*, muddy bottom feeder. The effluent containing organic wastes, exudates, dead and decayed matters are termed as detritus and it is not lost to the ecosystem, rather it serves as the source of mercury flow through detrivores like *Mugil cephalus*. The high level of mercury accumulation (3.92 $\mu g\ g^{-1}$) is mainly due to intake of detritus food, absorption by gill and intestine of skin. The excretory materials present in the bottom sediments are the route of recycling of mercury. Hence, *Mugil cephalus* forms an important item of food for man and it can become an indirect source of mercury entering the body, leading to toxic effects. The mercury accumulation in different food-chain organisms was found to be in the decreasing order from the source to water (*Effluent > Zooplankton > Phytoplankton > Sediment > Bivalve > Fish*). In the present study mullet showed 0.23 $\mu g\ g^{-1}$ (ppm) in the edible portion (tissue) which is lesser than the maximum permissible limit of mercury for human consumption prescribed as 0.5 ppm by USSR,

0.7ppm by Italy and 0.4 ppm by Japan (FAO, 1983). In the present study the bioaccumulation of mercury found to be safer level but due course the level may increase and cause several problems to the environment. Already there were reports on flourosis in the rural areas near by the industries. Further ground water turns useless for drinking and bathing, smoke released by the industries cause bad odour and affects the respiratory system also. During monsoon season, the effluents washed from all the industries and discharge in to the Uppanar and hence mass mortality of finfish and shellfish was noted. Therefore, Government of India should take necessary steps to control the pollutants released from all the industries in this area.

Acknowledgements

The authors are thankful to the Director, Centre of Advanced Study in Marine Biology, Annamalai University, Parangipettai for providing the facilities and encouragements during the study.

References

Boyden, C.R and M.G. Romeril (1974). A trace metal problem in pond oyster culture. *Mar. Pollut. Bull.*, 5: 74-78.

Brooks, R.R., B.J. Presley and I.R. Kalplan (1967). APDC-MIBK extraction system for the determination of trace metals in saline water by atomic absorption spectroscopy. *Talanta*, 14: 809–816.

Bryan, G.W. (1976). Some aspects of heavy metal tolerance in aquatic organisms. In: *Effects of Pollutants on Aquatic Organism*, (Ed.) Lockwood, A.P.M. Cambridge University Press, Cambridge, p. 7–34.

Chester, R. and M.J. Hughes (1967). A chemical technique for the separation of Ferro-manganese minerals, carbamate and absorbed trace elements from pelagic sediments. *Chem. Geol.*, 2: 249–263.

Cooper, J.J. (1983). Total mercury in fishes and selected biota in Lohotan reservoir, Navada. *Bull. Environ. Contom. Toxicol.*, 31: 9–17.

Davis, A.G. (1978). Pollution status with marine plankton. Part. II. Heavy metals. *Adv. Mar. Biol.*, 15: 381–508.

FAO (1983). *Compilation of Legal Limits for Hazardous Substance in Fish and Fishery Products*, pp. 102.

Judith E. Foulke (2000). Mercury in fish; cause for concern. *Sea Food Export J.*, 31(1): 123–134.

Knauer, G.A. and J.H. Martin (1973). Seasonal variation of cadmium, manganese, lead and zinc in water and phytoplankton in Monterey Bay, California. *Limnol. Oceanogr.*, 18(4): 597–604.

Patel, B. and J.P. Chandy (1988). Mercury in the biotic and abiotic matrices along Bombay coast. *Indian J. Mar. Sci.*, 17: 56–58.

Paul, Raj S. (1986). *Poll. Res.*, 5(2): 39–43

Paul, Raj S. (1987). *J. Environ. Biol.*, 8(2): 151–155.

Topping, G. (1973). Heavy metals in fish from Scottish water. *Aquaculture*, 1: 373–377.

Chapter 9

Toxic Effect of High Chromium Intake on the Human Body and Chromium Removal from Water with Low Cost Adsorbents

Jyotsna Lal and Kanchan Sachan

ABSTRACT

The heavy metal chromium is emitted from various sources for example electroplating industry, paint and pigment waste, cotton textiles dyeing, glass chrome leather tanning industries. Hexavalent form of Chromium is more toxic than trivalent form, Cr (VI) causes bronchial cancer in man. The maximum permissible limit of Cr in drinking water. As recommended by WHO is 0.05 mg/l. The industrial town of Kanpur has a many cotton textiles, chrome leather tanning, chemical manufacturing industries. A survey was conducted and found that in comparison to other industrial areas of India the ground water of this area contains a very high concentration Cr (VI). Most water treatment technologies suitable for the developed countries are expensive for towns and small industries of the developing countries because of limitation of funds and lack of skilled man power. There is a increasing search for low cost adsorbent material made from locally available waste material and plant substrata. Some low cost adsorbents selective for Cr (VI) have been tested in the laboratory by batch and column technique pH, concentration and flow rate on break-through capacity have been investigated.

Keywords: *Virons or carcinogenic metals, Chromium, Sawdust, Tealeaves carbon, Red mud rice husk carbon, Rice straw, Brick kiln ash, Flyash, Groundnut husk carbon, Baggase.*

Introduction

Carcinogen chromium is emitted from waste effluents of electroplating, paint, pigment waste, cotton textiles dyeing, glass, chrome leather tanning industries (Table 9.1) (Poonam Tyagi 1999). A survey was conducted and found that in comparison to other industrial areas of India the ground water of industrial town of Kanpur in Uttar Pradesh contains a very high concentration Cr(VI) (Figure 9.1) (GWQS/8/1996–97). The max permissible limit of Cr(VI) in drinking water, recommended by WHO is 0.05 mg/l. Hepato and nephrato toxic effects have been reported in rats (Meenakshi, 1989). Cr is highly toxic to fish and other aquatic life, causes changes in oxygen uptake and blood parameters in the fish *Sarotherodan mossambicus* (Peters) (Pamila, D., 1991).

Table 9.1: Chromium in the Groundwater of Industrial Areas of India

Area	*mg/L*	*Industry*
Rourkela, Bihar	0–0.01	Steel plant
Cochin, Kerala	0.008–0.019	Tanneries
Jharia, Bihar	0.01–0.09	Ore-mining
Himachal Pradesh	0.001–0.019	Ore-mining
Haryana		Bicycles
Punjab		Woollen garments

The virons or carcinogenic metals with their tendency for chelation are associated with vitamins; protein and nucleic acid. These metals ions may replace the (ZnII, MgII) from the normal enzyme system or nucleic acid and may alter the structure and function of genetic material resulting in the formation of cancer. It has been shown that specific metals are required for genetic regulation and wrong metal ions, even the essential metal ions in wrong concentrations can lead to errors resulting in metal

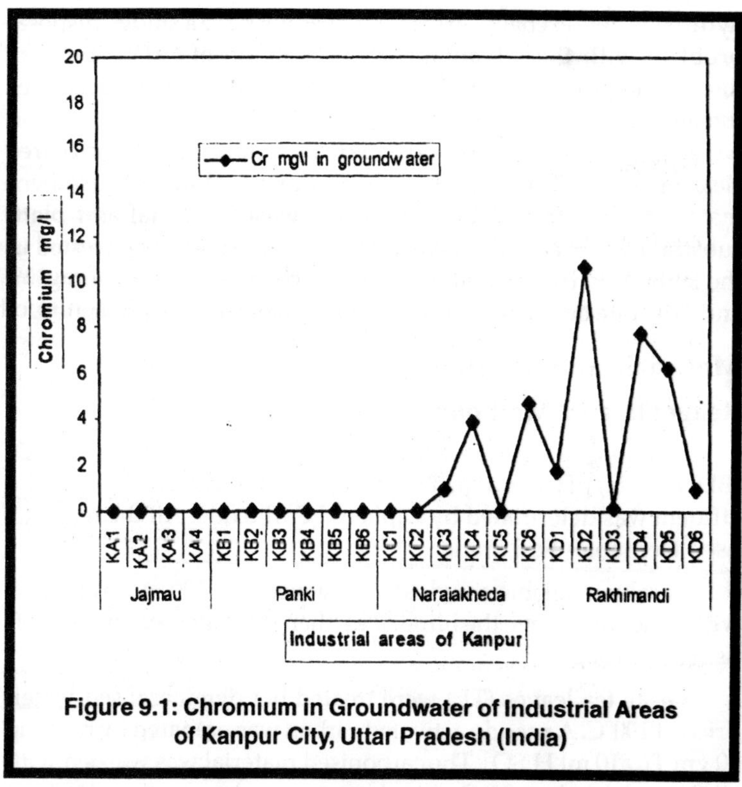

Figure 9.1: Chromium in Groundwater of Industrial Areas of Kanpur City, Uttar Pradesh (India)

carcinogenesis. The fashioned DNA molecule require only deoxy nucleotides and not ribonucleotide be incorporated into the nucleic acid. The ability of DNA polymerase 1 to distinguish between these two monomers types has been shown to depend on the nature or the metals used to activate the enzyme thus the presence of Mg(II) insure incorporation of deoxynucleotides, but Mn(II) and Cr(VI) permits the concurrent incorporation of ribonucleotides. Thus the presence of Mn(II) and Cr(VI) lead to incorporation of the wrong sugar (Ribonucleotides) in place of deoxynucleotides into the nucleic acid with both type of enzyme (RNA and DNA polymerase) carcinogenic metals are suppose to cause error (Chakrawarti, 2003).

Precipitation, ion-exchange, adsorption are technologies available for removing Cr(VI) from water. Reduction and precipitation of chromate with barium or lead, the major problem with this type of treatment is the disposal of precipitated chromium hydroxide. Ion-exchange treatment does not have a sludge disposal problem with the advantage of reclamation of Cr(VI) but has disadvantages of being expensive, poorly selective for chromate over foreign anions and a critical flow rate while the process of adsorption has an edge over other methods due to its simplicity and sludge free clean operation. There is an increasing search for low cost adsorbent material made from locally available waste material and plant substrata. Some adsorbents selective for Cr(VI) have been tested in the laboratory by batch and column technique, pH, concentration and flow rate on break-through capacity have been investigated.

Materials and Methods

Reagents and Methods

Solutions of Cr(VI) were prepared in tapwater using $K_2Cr_2O_7$ (BDH, India) pH was adjusted with HCl. Presence of Cr (VI) in the effluent was determined by Diphenyl Carboxide. All the reagents used were analytical grade.

Locally available sawdust was sieved (60–100 mesh) treated with water to remove the impurities, then the water was filtered off using cloth bag.

Waste tea leaves (TL) were treated hot demineralised water, dried at 100°C. A pseudo activated carbon was prepared by reacting 10 gm TL/10 ml H_2SO_4. The carbonised material was washed with DMW and dried at 100°C, sieved (60–100 mesh) gave the adsorbent waste tealeaves carbon. Rice straw was procured from the fodder of the local market cut 1 cm long size, soaked in water overnight, washed free of colored impurities and dried sieved (60–100 mesh). Adsorption capacity by column and batch technique, break-through capacity of Cr(VI) was determined by passing Cr(VI) solution through a glass column (i.d. 1.25 cm) packed with 5 gm adsorbent on a glasswool support. The flow rate was maintained at 125 ml/h. Batch studies were carried out in glass stoppered conical flasks containing 100 ml solution with different dosage and contact time in a mechanical shaker.

Results and Discussion

Some low cost adsorbents selective for Cr(VI) have been tested in the laboratory by batch and column technique pH, concentration and flow rate on break-through capacity have been investigated.

The author compared some adsorbents which are available through out the year. Human Hair was a poor adsorbent at pH range 1–5 removed 1.1 mg Cr/gm (Tan, 1985). Similarly Brick kiln ash showed only 50 per cent removal as compared to Fly ash at pH 1.3 from chromeplating wastewater as shown in Figure 9.2 (Rai, 1999). Bituminous coal at pH 2 removed 2.5mg/gm (Narayan, 1989). Adsorbent made from a mixture Fly ash–Chinaclay showed 100 per cent removal at pH 2 from solution 4.3mg Cr(VI)/L (Pandey, 1984).

Columns packed with Sawdust a natural inexpensive material, available everywhere can be utilized as low cost adsorbent for the removal of Cr(VI) from aqueous solution, the adsorption capacity was determined as 97.5 mg/5 gm or 19.5 mg/gm at pH~2. Break through capacity as a function of pH is plotted in Figure 9.3. Rice straw an inexpensive material can be utilized as low cost adsorbent for the removal of Cr(VI) from aqueous solution (Figure 9.4), the adsorption capacity was determined 25mg Cr/gm. Baggase is reported to remove 80.80 per cent at pH 6–7 from solution 3.5gm Cr(VI)/L (Rao, 1999). Tea leaves carbon (TLC) can play an important role in treatment processes of metal bearing industrial effluents, a cheap carbonaceous adsorbent made from waste tealeaves break through capacity 39.3 mg/m. In Figure 9.5, Cr(VI) removal as a function of pH and carbon dosage can be seen. The amount of TLC required for 100 per cent Cr(VI) removal 433 mg Cr(VI)/L carbon dosage 1200 mg/L at pH 1.5–2.0. The author compared TLC to Groundnut husk carbon GHC. It can be seen that, activated Groundnut husk carbon at pH range 0.5–1.7 showed 100 per cent removal from 10 mg Cr(VI)/L solution, carbon dosage 2400 mg/L in Figure 9.6 (Periswamy, 1991).

Presence of holocellulose and lignin on sawdust and rice straw can be explained for removal of Cr(VI) from aqueous solution. Sigworth and Smith 1972 have proposed the possibility of heavy metal precipitation on the carbon surface by nucleation as one of the pathways for heavy metal removal by activated carbon. The adsorption of Cr(III) on red mud is highly dependent on pH = 5–6 Cr(III) adsorption on red mud also depends upon the contact time, sorbent dose and initial concentration of Cr(III). Red mud showed 99.6 per cent Cr(III) removal from a solution containing 150 mg Cr (III)/100 ml (Figure 9.7) (Kumar Raj, 1999).

Iron hexamine modified sawdust removed 19.53 mg Cr(VI)/gm in presence of other ions in pH range 2–6 (Figure 9.8) (Singh, 1992).

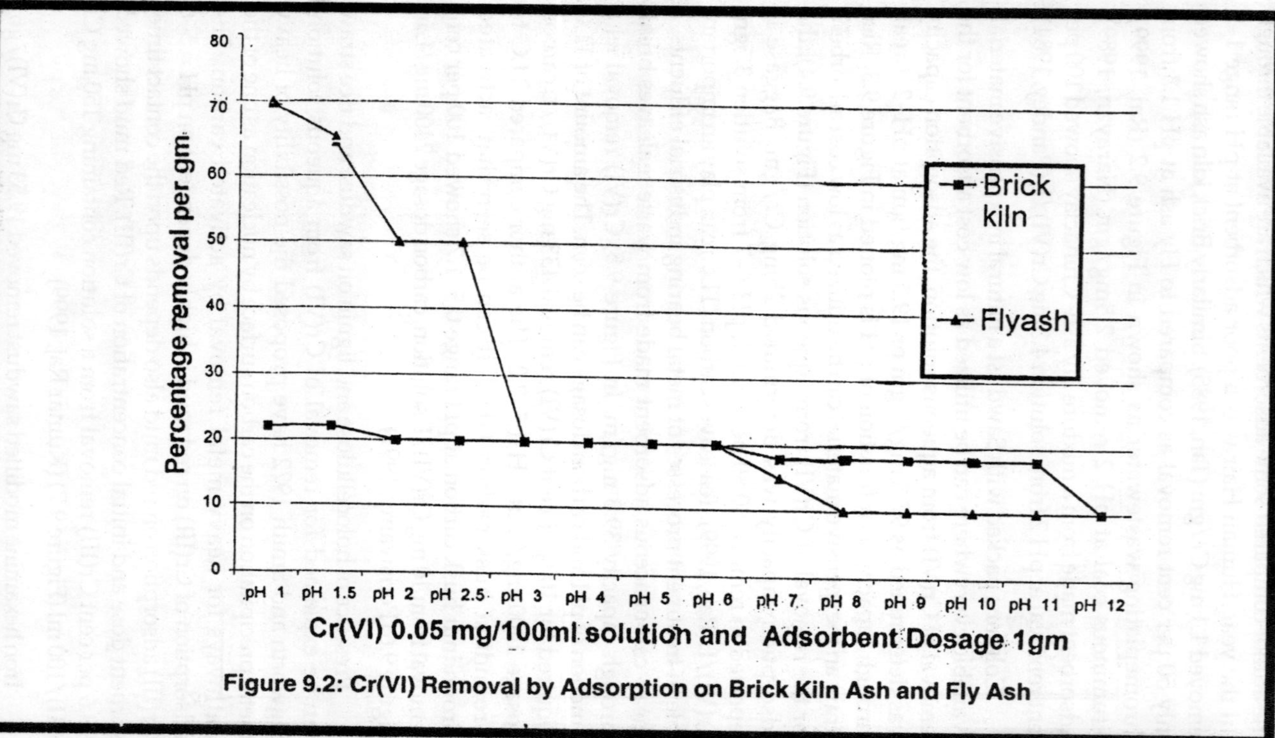

Figure 9.2: Cr(VI) Removal by Adsorption on Brick Kiln Ash and Fly Ash

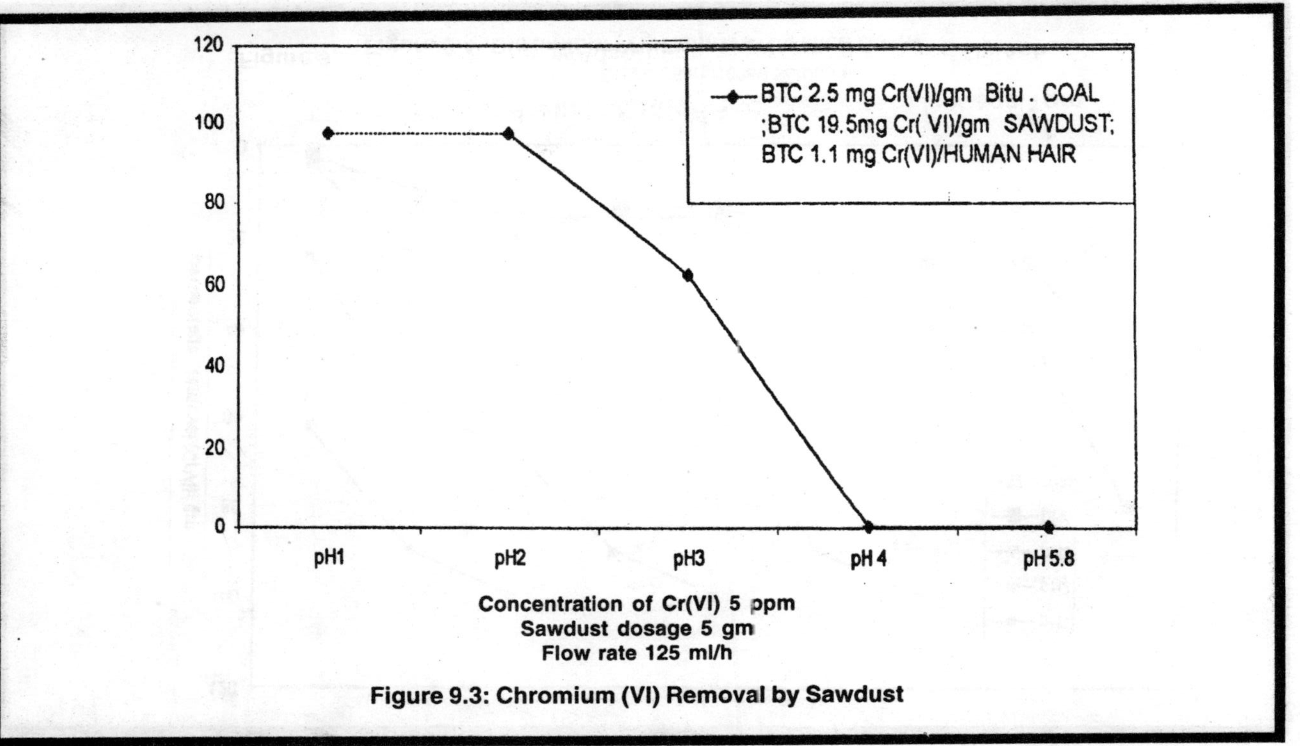

Figure 9.3: Chromium (VI) Removal by Sawdust

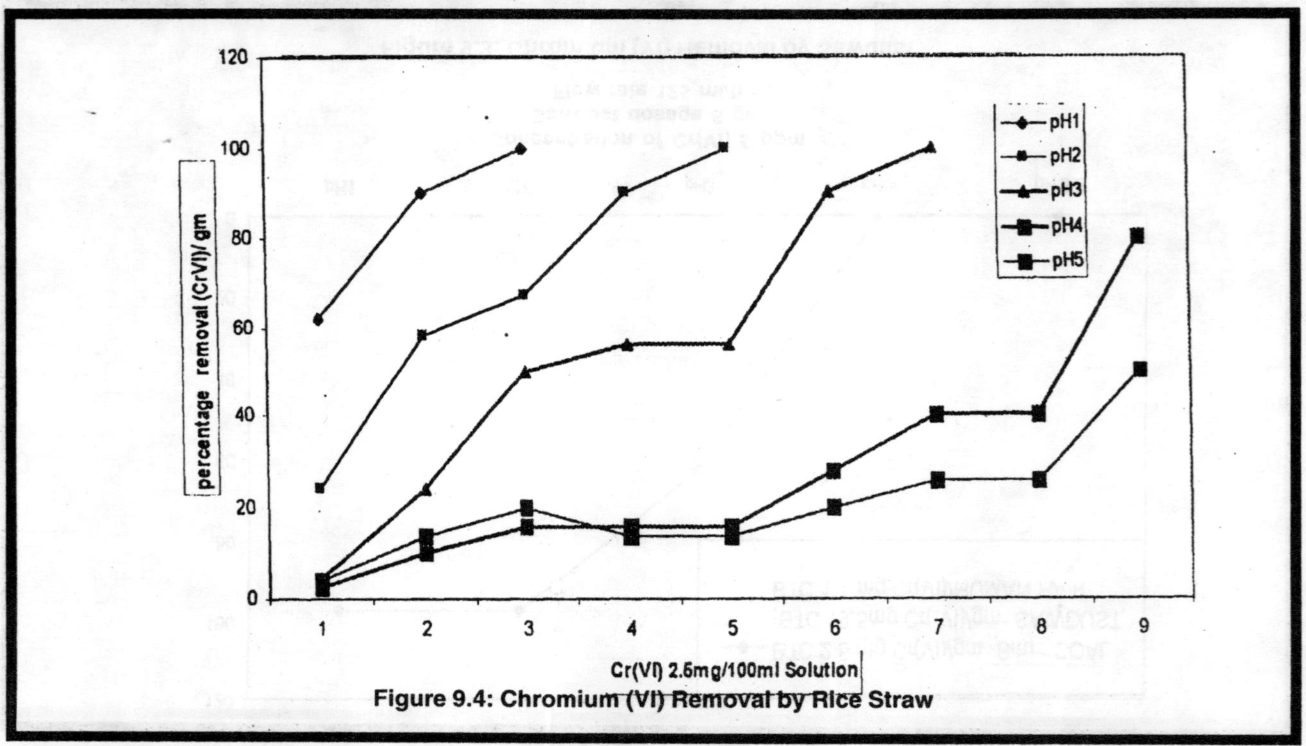

Figure 9.4: Chromium (VI) Removal by Rice Straw

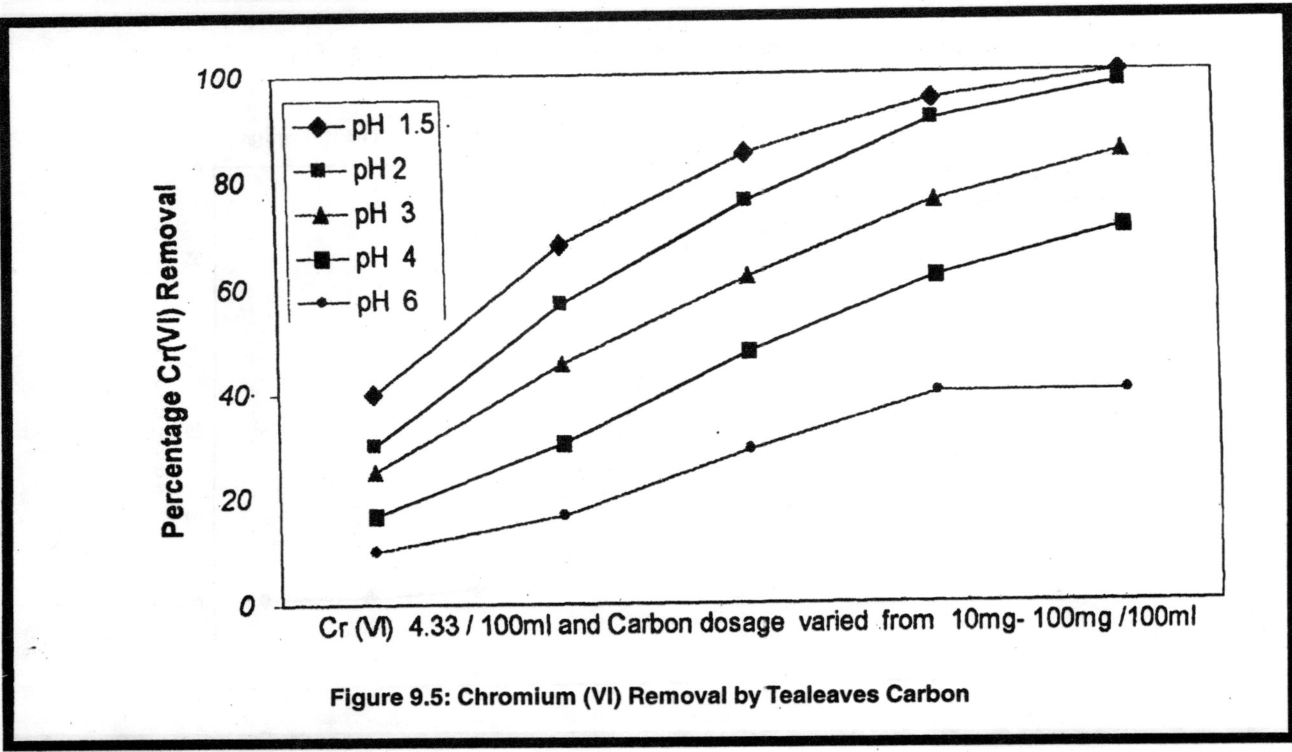

Figure 9.5: Chromium (VI) Removal by Tealeaves Carbon

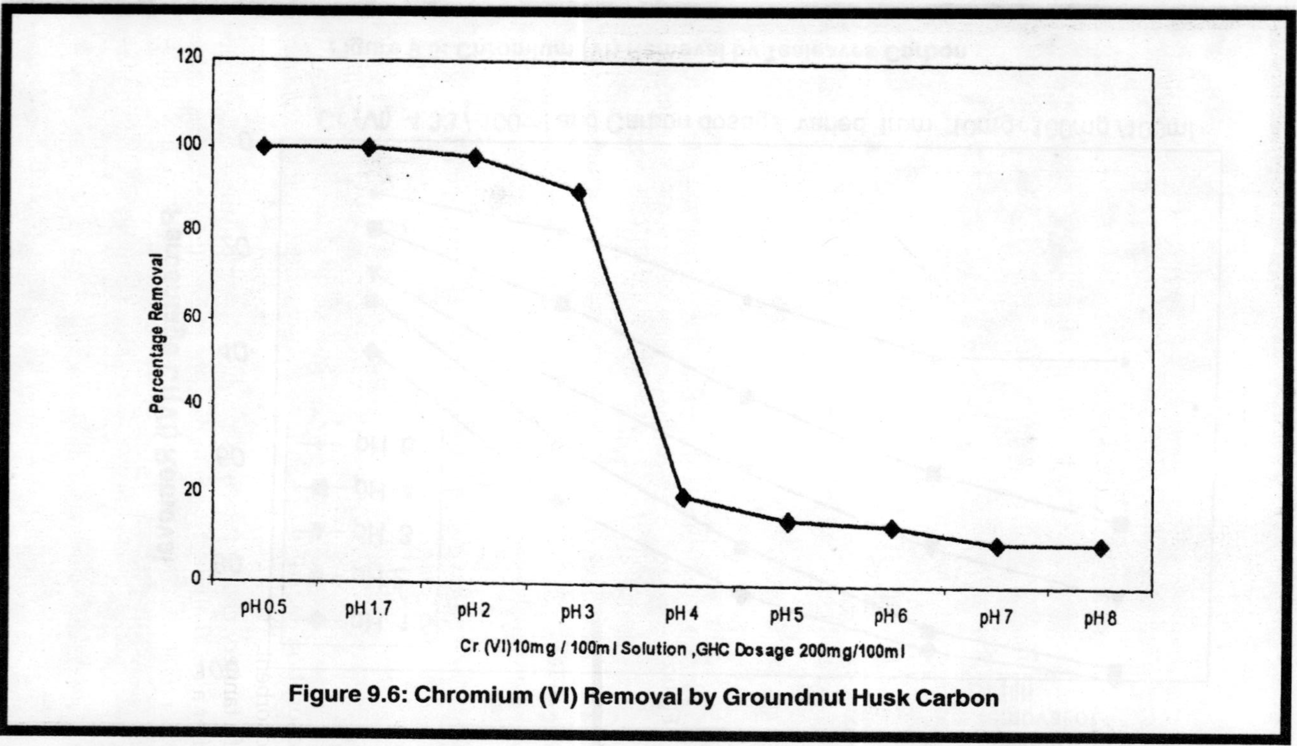

Figure 9.6: Chromium (VI) Removal by Groundnut Husk Carbon

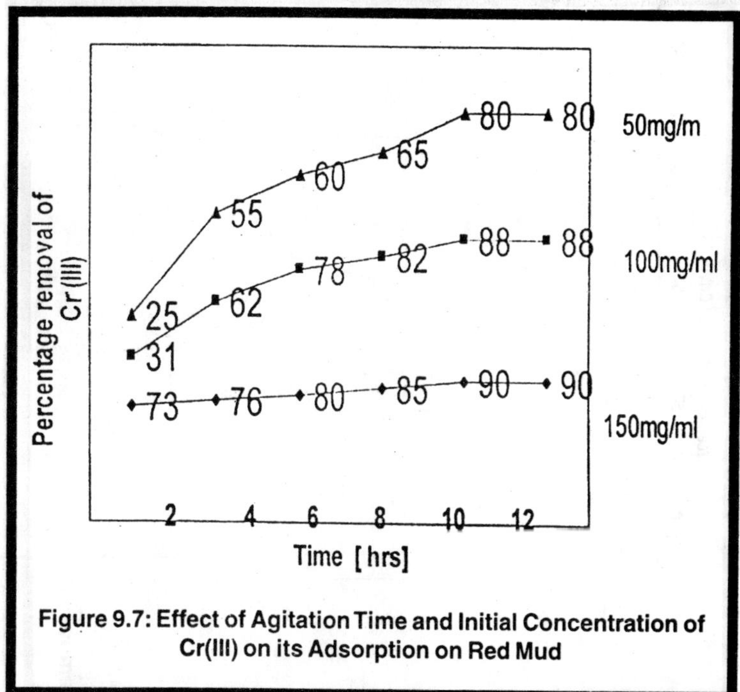

Figure 9.7: Effect of Agitation Time and Initial Concentration of Cr(III) on its Adsorption on Red Mud

Rice straw, Baggase and Sawdust are natural inexpensive materials, which are easily available and do not require chemical pre-treatment, exhibit exceptionally high degree of Cr(VI) adsorption, thus can be utilized for treatment of industrial waste effluents containing Cr(VI). Red mud may serve as economic and effectie fixing agent for Cr(III) removal from chromium laden wastewater. On the other hand Tea leaves carbon TLC and Groundnut husk carbon GHC are excellent cheap carbonaceous adsorbents showing high degree of Cr(VI) adsorption. Iron hexamine modified sawdust is another cheap adsorbent. These adsorbents can easily prepared in any laboratory of a tannery, chrome plating or paint industry. All these adsorbents offer a good alternative to expensive activated carbons and other water treatment technologies. This article has been limited to a study of adsorption capacities at different pH for some low cost absorbents. There is a need to describe the mechanisms responsible for removal of Cr(VI) from aqueous solution by low cost absorbents, Cr(VI) adsorption and reduction to Cr(III) by these adsorbents (Singh and Lal, 1992).

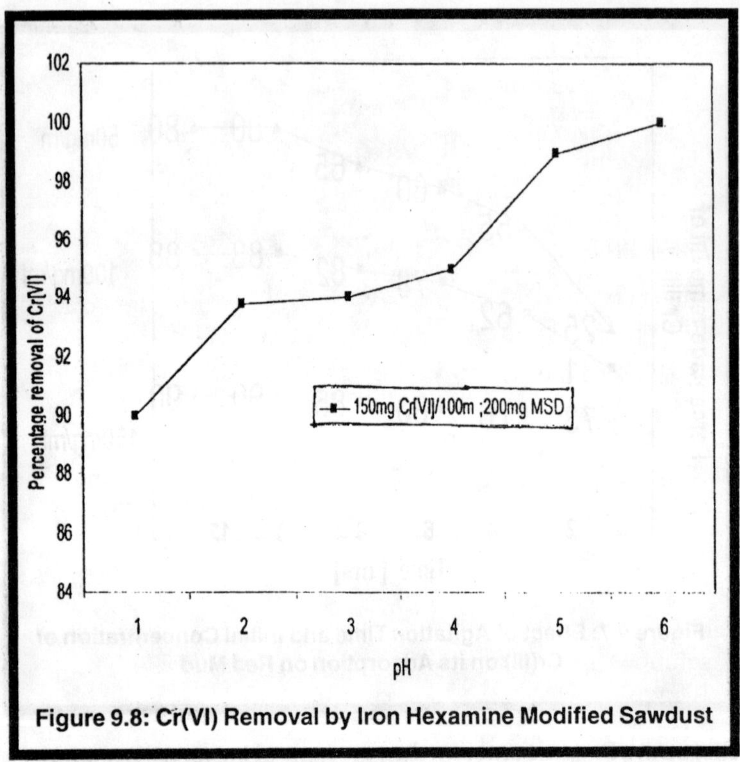

Figure 9.8: Cr(VI) Removal by Iron Hexamine Modified Sawdust

References

Chakravati, P.B. (2003). Chelation and cancer. *Science Letters*, 26, 11 and 12: 295.

Meenakshi, C.E. (1989). Comparative toxicity of trivalent and hexavalent Cr in Rats. *IJEH*, 31: 250–256.

Deo, Nandita and Mansoor Ali (1992). Optimisation of new low cost adsorbent in removal of Cr(VI) from industrial wastewater. *IJEP*, 12: 828–834.

GWQS/8/1996–97. Groundwater Quality Series, Groundwater quality in Kanpur: Status, Sources and Control Measures.

Narayan, N. and K.A. Abburi (1989). Study of removal of Cr(VI) by adsorption on bituminous coal. *IJEH*, 31: 304.

Pamila, D. (1991). Toxic effects of Cr on *Sarotherodan mossambicus* (Peters). *IJEH*, 33: 218–224.

Pandey, K.K. (1984). Flyash-Chinaclay for removal of Cr(VI) from aqueous solutions. *Ind. J. Chem.*, 234: 514.

Periswamy, K. (1991). Study on Cr(VI) removal by activated groundnut huskcarbon. *IJEH*, 33: 433–439.

Rai, A.K. and Surendra Kumar (1999). Treatment of Cr bearing wastewater by adsorption on Brick Kiln ash and Flyash. *Ind. J. Env. Hlth.*, 41: 65–73.

Rajkumar (1999). Removal of trivalent Cr–Red mud. *IJEP*, 21(2): 97–100.

Shantha, G.M. (1993). Adsorption studies on Cr(VI) removal–by activated carbon derived from Baggase. *J. Env. Sci. Hlth.*, 28(A): 2263–2280.

Sigworth, E.A. and J.A. Smith (1972). Activated carbon. *J.A.W.W.A.*, 64: 386.

Singh, D.K. (1990). Removal of Cr(VI) from aqueous solutions using low cost adsorbent. *JIE(I)–CH Eng.*, 70: 90.

Singh, D.K. and J. Lal (1992). Removal of Cr(VI) from aqueous solutions using waste tea leaves carbon. *Ind. J. Env. Hlth.*, 34: 108.

Singh, D.K. (1992). Removal of toxic metal ions–modified adsorbent. *Poll. Res.*, 11: 187.

Tan, T.C., Chia, C.K. and C.K. Teo (1985). Uptake of metal ions by chemically treated human hair. *Water Res.*, 19: 157.

Chapter 10

Metal Ion Stabilization of Alpha Amylase Activity in *Opuntia vulgaris*

B. Nabi and K. Srikumar

ABSTRACT

The present studies indicate the existence of two isoforms T_{50} and T_{90} of amylase activity which were independently capable of hydrolyzing the starch substrate at 50°C and 90°C respectively has been detected in the cladode of Opuntia vulgaris. The enzyme shows the pH optima at 8.5. The T_{90} isoform was found to be stabilized by Mg^{2+}, Ca^{2+} and less affected by Mn^{2+}. Metal stabilization of enzyme activity possibly offers an additional strategy in plants to grow in arid habitats.

***Keywords:** Amylase, Thermostable, Opuntia vulgaris, Starch, Metal ions.*

Introduction

Alpha-amylase (EC 3.2.1.1, 1, 4-alpha-D-glucan glucano hydrolase) hydrolyses starch, glycogen, and related polysaccharides by randomly cleaving internal alpha-1,4-glucosidic linkages. Alpha-amylase is an important industrial enzyme. It is well distributed in various bacteria, fungi, plants, and animals and has a major role in utilization of polysaccharides. It can be used for removal of starch sizing from textiles, additive in detergents, the liquefaction of starch, and the proper formation of dextrin in baking. Most of the thermophilic amylase require Ca^{2+} for their thermostability (Long-Liu Lin, 1998), and show a wide range of substrate specificities towards different substrates. A variety of thermophilic bacteria, for example *Bacillus stearothermophilus* (Wind, 1994) and *Thermococcus profundus* (Chung, 1995), produce amylases. The advantages of using thermostable alpha-amylases in industrial processes include the decreased risk of contamination, the increased diffusion rate and the decreased cost of external cooling. So far several reports for the thermophilic enzymes are from the bacterial origin (Tsvetkov, 1989; Freer, 1993; Manning, 1961), but no reports on thermophilic amylase from *Opuntia vulgaris* have been reported. Studies carried out in our laboratory identified the presence of an amylase enzyme activity in the xerophyte, *Opuntia vulgaris.* The pH optimum of the enzyme was 8.5 and it was found to be active in the temperature range 50–90°C. This enzyme activity was found to be further stabilized by specific divalent cations. We report the presence of a thermostable amylase in *Opuntia vulgaris* potentially useful for industrial applications.

Materials and Methods

Chemicals

Starch, Phenylmethanesulphonyl fluoride (PMSF), $MgCl_2$, $CaCl_2$, $MnCl_2$, KH_2PO_4 and Na_2HPO_4. All the other chemicals used were of analytical grade obtained from manufacturers in India. Glass distilled water was used for the preparation of all reagents.

Enzyme Extraction

Opuntia vulgaris cladode homogenate 20 per cent (w/v) was prepared in 0.1M KH_2PO_4/Na_2HPO_4 buffer, pH 7.0, containing 1mM PMSF. The extract was filtered through cheese cloth and the filtrate was centrifuged at 10,000 x g for 15 min in a refrigerated centrifuge to obtain a clear supernatant that was used as the enzyme extract.

Enzyme Assay

Alpha-amylase activity was assayed using a previously described method Hemy and Chiamori (1964) by measuring the amount of reducing sugar released during the reaction employing starch as the substrate. The reaction mixture contained 1.0 ml soluble starch 1.075 per cent, 1.5 ml of 0.1 M KH_2PO_4/Na_2HPO_4 buffer, pH 7.0, and 100 µl of the enzyme extract. The reaction was incubated for 30 min, stopped by the addition of 0.5 ml 2/3 N H_2SO_4 and 0.5 ml 10 per cent sodium tungstate, and centrifuged at 5000 × g for 15 min in a refrigerated centrifuge, 1.0 ml of supernatant was taken and 1.0 ml of alkaline copper reagent was added to the reaction sample and the sample was heated in a boiling water bath for 6 min. 1.0 ml of phosphomolybdic acid was then added and incubation of the sample continued in water bath for an additional 2 min. The blue colour developed in the sample was read at 650 nm employing a UV–visible single beam spectrophotometer.

Effect of Metal Ions and on Enzyme Activity

Studies related to the effect of metal ions on the amylase enzyme activity was carried out employing divalent cations in their chloride forms at different temperatures (30°C–90°C) in KH_2PO_4/Na_2HPO_4 buffer at pH 7.0, incubating the samples each time for 30 min. The metal ions were used at 1, 5 and 10 mM concentration in independent reactions. The activity of amylase assayed in the absence of metal ions served as the control.

Effect of pH on Enzyme Activity

The enzyme activity was measured by incubating the reaction mixture at selected pH as described earlier for the enzyme assay. Measurement of pH stability, were carried out at 37°C for 30 min employing the enzyme in different buffers. The buffers used were sodium acetate buffer (pH 4.5–6.5), KH_2PO_4/Na_2HPO_4 buffer (pH 7.0–8.0) and Tris/HCl buffer (pH 8.5–9.5).

Results and Discussion

The temperature profile of alpha amylase activity in *Opuntia vulgaris* cladode yielded two temperature optima, one at 50°C (T_{50}), and the other at 90°C (T_{90}) results not shown. The activity of amylase was found to be optimum at pH 8.5 (Figure 10.1.). The enzyme activity was influenced by divalent cations Mg^{2+}, Ca^{2+} and Mn^{2+} used in their chloride form in the temperature range 30°C–90°C (Table 10.1). Ca^{2+} is important for the amylases obtained from the mesophilic sources. The enzyme was not stimulated by Ca^{2+} (Paquet, 1991). Each divalent cation influenced the amylase activity dependent upon

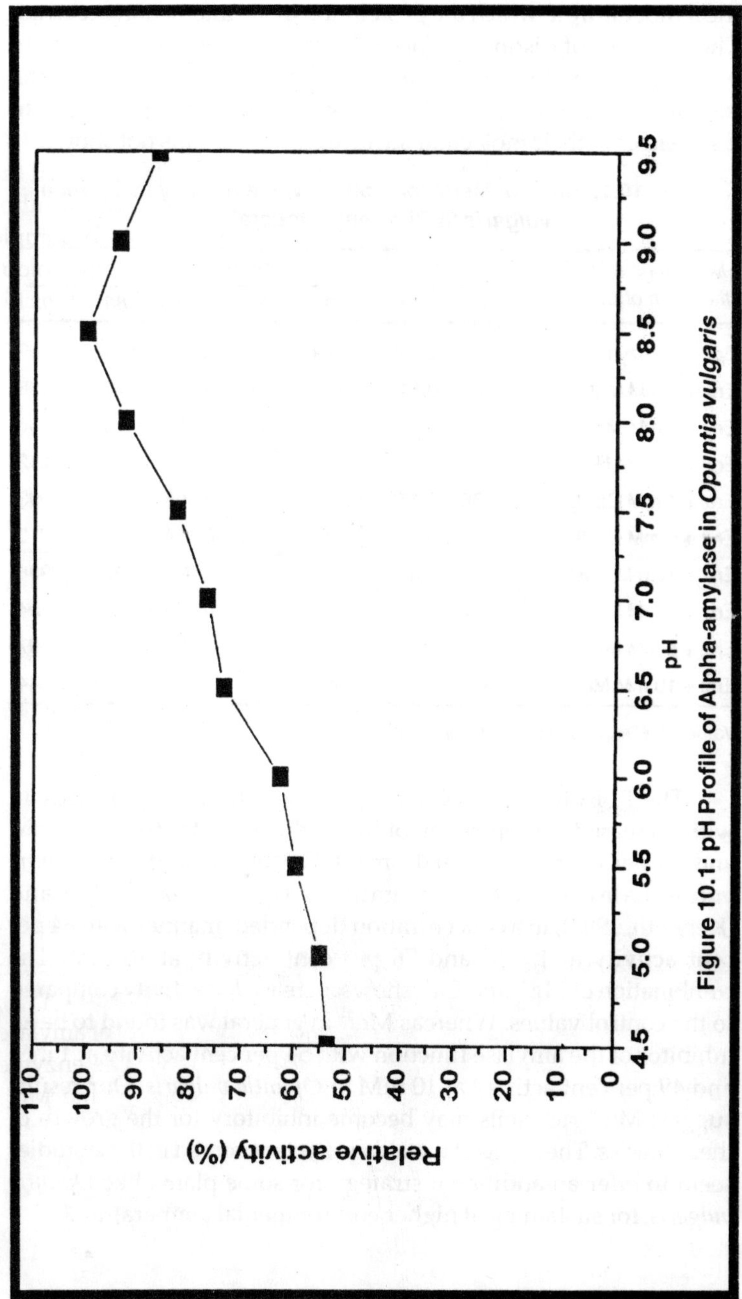

Figure 10.1: pH Profile of Alpha-amylase in *Opuntia vulgaris*

the temperature at which the protein-ion metal interaction occurred. The presence of distinct isoforms (T_{50} and T_{90}) of amylase activity was observed in the *Opuntia vulgaris* which were separable in the ion exchange column as weakly and strongly anionic species due to differences in their molecular surface charges results not shown.

Table 10.1: Effect of Metal Ions on Amylase Activity in *Opuntia vulgaris* at Different Temperature

Metal Ions Added in the Form of Chloride	*Temperature (°C)*						
	30	*40*	*50*	*60*	*70*	*80*	*90*
Control Enzyme	0.454	0.842	0.879	0.743	0.638	0.784	1.309
Enz + 1 mM $MgCl_2$	0.223	0.516	0.938	0.550	0.329	0.470	1.138
Enz + 5 mM $MgCl_2$	0.214	0.315	1.109	0.571	0.729	0.601	1.150
Enz + 10 mM $MgCl_2$	0.304	0.327	0.481	0.343	0.536	0.426	1.212
Enz + 1 mM $CaCl_2$	0.366	0.173	0.782	0.419	0.568	0.762	1.093
Enz + 5 mM $CaCl_2$	0.412	0.449	0.784	0.515	0.564	0.794	1.015
Enz + 10 mM $CaCl_2$	0.470	0.872	0.838	0.670	0.715	0.854	0.996
Enz + 1 mM $MnCl_2$	0.097	0.389	0.854	0.423	0.568	0.872	0.704
Enz + 5 mM $MnCl_2$	0.373	0.315	0.932	0.207	0.631	0.587	0.697
Enz + 10 mM $MnCl_2$	0.323	0.646	0.858	0.571	0.330	0.711	0.644

Values are expressed in activity (U/ml)

The T_{90} isoform at 90°C for Mg^{2+} shows the increase in activity with increased concentration of Mg^{2+} (87 per cent activity at 1 mM and 93 per cent activity at 10 mM). Ca^{2+} influenced the *Opuntia vulgaris* amylase activity negatively (Jei-Fu Shah, 1995) and (Kenneth, 1993), in a concentration dependent manner with 84 per cent activity at 1 mM and 76 per cent activity at 10 mM. The combination of Mg^{2+} and Ca^{2+} shows increased in activity compared to the control values. Whereas Mn^{2+} in general was found to be an inhibitor of the amylase function with 54 per cent activity at 1 mM and 49 per cent activity at 10 mM in *Opuntia vulgaris.* Our results suggest Mn^{2+} rich soils may become inhibitory for the growth of these species. The divalent metal ion effects observed in these studies seem to offer an additional strategy for some plants like *Opuntia vulgaris,* for sustaining at higher environmental temperature.

References

Chung, Y.C., Kobayashi, T., Kanai, H., Akiba, T. and T. Kudo (1995). *Appl. Environ. Microbiol.*, 61: 1502–1506.

Freer, S. (1993). Purification and characterization of extracellular alpha amylase from *Streptococcus bovis* JB1. *Appl. Environ. Microbiol.*, 59: 1398–1402.

Henry, R.J. and N. Chiamori (1960). *Clin. Chem.*, 6: 434.

Henry, R.J. (1964). *Clin. Chem.*, 471.

Jei-Fu Shah, Fu-Pang Lin, Su-Chiu Chen, and Hsing-Chen Chen (1995). Purification and properties of an extracellular alpha-amylase from *Thermus sp. Bot. Bull. Acad. Sin.*, 36: 195–200.

Kenneth A. Laderman, Bradley R. Davis, Henry C. Krutzsch, Marc S. Lewis, Griko, Y.V., Peter L. Privalov and Christian B. Anfinsen (1993). The purification and characterization of an extremely thermostable alpha amylase from the hyperthermophilic Archaebacterium *Pyrococcus furiosus. J. Biol. Chem.*, 268: 24394–24401.

Long-Liu Lin, Charng-Cherng Chyau and Wen-Hwei Hsu (1998). Production and properties of a raw starch–degrading amylase from the thermophilic and alkaliphilic *Bacillus sp.* TS–23. *Biotechnol. Appl. Biochem.*, 28: 61–68.

Manning, G.B. and L.L. Campbell (1961). Thermostable amylase of *Bacillus stearothermophilus. J. Biol. Chem.*, 236: 2952–2957.

Paquet, V., Croux, C., Goma, G. and P. Soucaille (1991). Purification and characterization of extracellular alpha amylase from *Clostridium acetobutylicum* ATTCC 824. *Appl. Environ. Microbiol.*, 57: 212–218.

Tsvetkov, V. and E. Emanuilova (1989). Purification and properties of heat stable alpha amylase from Bacillus brevis. *Appl. Microbiol. Biotechnol.*, 31: 246–248.

Wind, R.D., Buitellar, R.M., Eggink, G., Huizing, H.J. and L. Dijkhuizen (1994). Characterization of a new *Bacillus stearothermophilus* isolate a highly thermostable alpha amylase producing strain. *Appl. Microbiol. Biotechnol.*, 41: 155–162.

Chapter 11

Phytotoxicity of Copper and Lead in Leafy Vegetable Amaranth (*Amaranthus oleraceous*)

M.I.S. Saggoo, N. Verma and P. Sharma

ABSTRACT

Presently experiments were conducted to study the effect of heavy metals copper and lead on growth of amaranth, a common leafy vegetable of summer months. The plants were grown from seeds in pots containing soil amended with metal salts to make the final concentrations of 10, 50, 100 and 500 ppm of copper or lead. Both the metals have toxic effect on seed germination but no effect was observed on morphological parameters of mature plants. Plants from metal rich soils showed delay in flower initiation with no significant toxicity in pollen. The leaves showed the uptake of copper which is on the risk side of the prescribed safety norms.

Keywords: *Phytotoxicity, Copper, Lead, Amaranth.*

Introduction

Urbanization, industrialization and anthropogenic activities have resulted in recycling pattern of various elements in the environment. Heavy metals are very significant ingredient of the environment that influences flora and fauna in many ways. Many heavy metals are essential for growth and metabolism at low concentration yet are toxic in excess. Plants accumulate both essential as well non-essential elements in their tissues. Toxicity of a number of heavy metals has been established by many workers (Antonovics *et al.*, 1971; Chaphekar and Shetye, 1988; Bhowmik and Sharma, 1999). The agronomic efficiency and potential hazards of heavy metal rich soils have to be assessed preferably by direct experimentation. The present experiment, therefore, was aimed to study the effect of two common heavy metals, copper and lead (one essential and other non-essential) on growth of leafy vegetable amaranth vernacular *chalai*. It was hoped to determine whether the edible part contained metal concentration that would impose health risk.

Material and Methods

To study the effect of copper and lead accumulation on characteristics of various crop plants pot experiment was performed in a greenhouse at Punjabi University, Patiala, India. Plants were raised in earthen pots (dia 9") filled with normal garden soil amended with known quantities of $CuSO_4$ as source of copper and $Pb(NO_3)_2$ as source of lead. For preparing 10 ppm of Cu in soil 112.96 mg of $CuSO_4$ was added in solution form in 3 kg or soil per pot. For same concentration of Pb 74.95 mg or lead nitrate was added in 3 kg of soil per pot. Four different concentrations of these salts were mixed in soil to make the final concentration of metals as 10, 50, 100, 500 ppm in soil of each pot. Garden soil without mixture of any salt served as the control. Whole experiment was triplicated.

Amaranth (*Amaranthus oleraceous.* Willd. cv. *haritbhaji*) plants were raised from seeds. Observations were made on morphological as well as reproductive characters from plants grown in pots. Pollen viability was calculated following TTC method (Shivanna and Rangaswamy, 1992).

The amount of Cu and Pb accumulated by plants was estimated by carrying out acid digestion and then analyzing the digested samples through Atomic Absorption Spectrophotometer (APHA, 1995).

Results and Discussion

Amaranth, an erect succulent pot herb with large oblong leaves, is cultivated from April to September throughout India. The leaves and tender shoots are ready for harvesting in 30–40 days. The plants particularly the green parts are good source of proteins and iron.

Effect on Early Growth

Effect of $CuSO_4$ and $Pb(NO_3)_2$ on early growth characters of *Amaranthus* was studied using petri-plate method. Both these salts showed deleterious effects on seed germination, which showed reduction with increase in concentration of copper and lead. Earlier studies showed different salts of lead to reduce seed germination in *Brassica* (Wong *et al.*, 1984), *Sorgham* (Peng and Guo, 1993), *Vigna* (Subramani *et al.*, 1997), etc. Copper sulphate (250 mg/L) on the other hand was found to be completely lethal for seed germination in *Vigna radiata* (Saravanan *et al.*, 1996).

When the length of root and shoot of the seedling was measured separately a decrease in seedling size with increase in concentration of copper was quite evident (Table 11.1). The average fresh weight of root and shoot in the control seedlings was 2.19 and 21,48 mg, respectively. A clear toxicity was observed in the seedlings of seeds treated with copper. In case of Pb treatment, the seedlings did not show much deviation from the control seedlings. Reduction in the growth of seedling roots under the influence of Pb has been reported earlier (Zheng and Chen, 1989; Hsu and Chou, 1992; Ochoa *et al.*, 1992, Salim *et al.*, 1995).

In the copper treated seedlings the absolute water content varied from 9.76 to 13.79 as compared to 14.21 in the control. Lead treatment increased the absolute water content to 19.86 (50 ppm Pb treatment). In copper treated seeds seedlings showed non-significant variation in percent dry matter. Percent dry matter showed reduction as compared to the control in case of Pb treated seedlings (Table 10.1). Higher amount of Cu and Pb reduced growth and dry matter in other crops (Laura *et al.*, 1993; Lanaras *et al.*, 1993). However, Izzo *et al.* (1992) recorded increase in dry weight of the roots of wheat exposed to lead.

Effect on Morphology

Amaranthus plants in the control pots grew up to 40 cm with an average height of 32.41 cm and leaf size 2.7 cm (Table 11.2). In copper amended soils there was an overall reduction in height of the plants than that of the control. Maximum reduction in size was observed in the plants raised on 500 ppm of Cu in soil. Similar toxic effect of copper has been reported in vegetables (Frank *et al.*, 1976; Hara and Sonoda, 1979), The analysis of correlation coefficient between plant

Table 11.1: Early Growth Characteristics of Amaranth Seedlings Raised on Different Concentrations of Copper and Lead

Treatment	*Seed Germination*	*Root Length Mean ± SD (cm)*	*Shoot Length Mean ± SD (cm)*	*Fresh Weight (mg)*		*Fresh Weight (mg)*		*Absolute Water Content*	*Dry Matter (%)*
				Root	*Shoot*	*Root*	*Shoot*		
Control	94.20	1.81 ± 0.89	4.07 ± 0.94	2.19	21.48	0.95	8.51	14.21	39.97
Copper									
10 ppm	88.18	1.42 ± 0.58	4.74 ± 1.04	1.51	17.61	0.71	4.62	13.79	27.88
50 ppm	85.56	1.62 ± 0.92	3.77 ± 0.92	1.76	13.83	0.82	5.01	9.76	37.40
100 ppm	85.05	1.21 ± 0.79	2.74 ± 0.54	1.82	15.01	0.91	6.03	9.89	41.24
500 ppm	76.90	0/77 ± 0.47	3.77 ± 0.81	1.96	18.83	1.01	7.02	12.76	38.62
Lead									
10 ppm	93.60	1.71 ± 0.99	3.59 ± 0.57	1.52	11.54	0.42	2.91	9.73	25.50
50 ppm	88.36	1.83 ± 1.05	4.41 ± 0.81	1.96	21.42	0.51	3.01	19.86	15.06
100 ppm	86.48	2.02 ± 0.94	4.46 ± 1.18	2.01	21.81	0.63	4.02	19.17	19.52
500 ppm	84.16	2.32 ± 0.96	4.80 ± 0.88	2.19	22.42	0.71	6.03	17.87	27.39

height and amount of copper added to soil revealed that there existed a negative but significant correlation ($r = -0.704$, $P = 0.01$). However, the correlation was negative and non-significant between plant height and copper accumulated in plants ($r = -0.494$).

Table 11.2: Characteristics of Amaranth Raised on Soil Amended with Different Concentrations of Copper and Lead

ppm of Metal Added to Soil	*Plant Height (cm)*	*Leaf Size (cm)*	*Days to Heading*	*Pollen Fertility (%)*	*Duration of Pollen Viability (h)*
Control	32.41	2.70	72	95.62 ± 3.02	38
Copper					
10 ppm	22.42	2.91	59	97.25 ± 2.05	29
50 ppm	28.27	2.44	62	98.21 ± 1.07	27
100 ppm	22.07	2.73	69	99.73 ± 0.36	30
500 ppm	18.32	3.11	71	99.86 ± 0.18	32
Lead					
10 ppm	35.32	5.15	61	96.87 ± 2.54	33
50 ppm	30.73	3.91	63	93.49 ± 2.48	30
100 ppm	37.34	4.92	69	98.12 ± 2.14	34
500 ppm	40.28	6.91	72	97.48 ± 0.98	36

Interestingly, the plants from the pots with lead amended soil were taller than that of the control (Table 11.2). Same trend has been observed for the size of leaves. Presence of lead in soil was shown to be responsible for reduced growth of various plants like sunflower (Kastori *et al.*, 1988), pulses (Prasad and Devi, 1990) and wheat (Lanaras *et al.*, 1993). The present observation of absence of any toxic response in the plants can be attributed to the non-absorption and non-accumulation of lead in the plants (Table 11.2).

Effect on Reproductive Characters

Days to Heading

Toxicity of the metals was, however, visible in reproductive characters. The plants from 10 ppm Cu were the first to reach flowering (50 per cent plants showed flowering on 59th day of sowing). Overall the plants from metal amended soils showed dose dependent delay in the flower initiation (Table 11.2). This observation is line with Sheoran and Singh (1996) who recorded delay in flower initiation in wheat due to lead and cadmium. Coefficient of correlation revealed positive correlation between day to heading and concentration of metal in soil.

Pollen Fertility

The presence of Cu and Pb in the soil had no significant effect on the pollen fertility of plants growing over it. Here no trend was evident (Table 11.2). Earlier studies have shown pollution related pollen sterility in different plants (Sharma, 1982; Saggoo and Kumari, 1994; Zhang, 1994). Pollen in *Amaranth* remained viable for 38 hrs after anthesis as studied through tetrazolium chloride test. When such study was performed on plants grown on Cu or Pb amended soils there was reduction in duration of pollen viability. Minimum duration of pollen viability of 27 h and 30 h was observed in plants grown over 50 ppm Cu and 50 ppm of Pb, respectively. Overall the reduction in duration of pollen viability was negatively correlated with metal in soil but was non-significant.

Metal Uptake

Amaranth plants raised on metal amended soil showed more uptake of copper by leaves than roots (Table 11.3). Maximum Cu accumulation in roots was observed in the plants from 50 ppm set (66.54 µg/g dry wt). At higher levels Cu uptake was reduced with no uptake in the roots of the plants of 500 ppm set. In case of the leafy parts increase in concentration of Cu was dose dependent. Plants from control pots too showed 9.23 µg/g dry wt of Cu in leaves.

Table 11.3: Accumulation of Cu in Amaranth Plants Raised in Copper Amended Soils

ppm Cu Added to Soil	*Cu µg/g dry wt.*	
	Root	*Leaf*
0	ND	9.23
10	48.80	48.80
50	66.54	687.64
100	22.18	745.31
500	20.71	754.18

Interestingly, Pb accumulation was not detected in the roots as well as leaves of amaranth plants from different concentrations of Pb in soil except 10 ppm where 14.90 µg/g dry wt. of Pb was recorded in leaves.

Food plants serve as an exposure pathway for many heavy metals. In the present case the edible part or amaranth *i.e.*, leaves have shown accumulation of 48.8 to 754.18 µg/g dry wt. of copper while no uptake of lead. Copper is essential element and 2–5 mg of intake is recommended for growth while more than 15 mg proved to be toxic (WHO, 1972). Since the level of Cu in the leaves is on the risk side it can be recommended that the vegetable from Cu rich soil should be avoided. While this can't be said about Pb as no accumulation of lead was observed in the leaves.

Acknowledgements

The facilities provided by Head, Department of Botany, Punjabi University, Patiala and Coordinator, DRS (UGC) are gratefully acknowledged.

References

Antonovics, J., Bradshaw, A.D. and R.G. Turner (1971). Heavy metal tolerance in plants. *Adv. Ecol. Res.*, 7: 1–85.

APHA (1995). *Standard Methods for Examination of Water and Wastewater*, 19th edition. American Public Health Association. Washington, D.C.

Bhowmik, N. and A. Sharma (1999). Lead Toxicity in Plant Systems: Some Aspects. *The Nucleus*, 131–181.

Chaphek, S.B. and R.P. Shetye (1988). Heavy metal toxicity and environment. In: *Environmental Issues and Researches in India*, (Eds.) S.K. Agarwal and R.K. Gerg, p. 115–128.

Dos Santos, C.M., Neto, M.M. P.M., De Varennes, A. and M.A.C. Fragoso (1993). Some effects of different levels of lead on Berseem. *Developments in Plant and Soil Sciences*, 53: 517–521.

Frank, R., Ishida, K. and P. Suda (1976). Metals in agricultural soils of Ontario. *Can. J. Soil Sci.*, 56: 181–196.

Hara, T. and Y. Sonoda (1979). Comparison of the toxicity of heavy metals to cabbage growth. *Plant Soil*, 51 (1): 127–133.

Hsu, F.H. and C.H. Chou (1992). Inhibitory effects of heavy metals on seed germination and seedling growth of *Miscanthus* species. *Botanical Bull. Acad. Sinica,* 33: 335–342.

Izzo, R.I., D'Alessio, D. and N.F. Izzo (1992). Chemical composition of crop production as a consequence of position in the crop order I Accumulation and distribution of main nutrients in *Triticum aestivum. Agrochimica* (Italy), 36: 193–203.

Kastori, R., Plesnicar, M., Sakac, Z., Pankovic, D. and I. Arsenijevic-Maksimovi (1998). Effect of excess lead on sunflower growth and photosynthesis. *J. Plant Nutri.,* 21: 75–85.

Lanaras, T., Moustakas. M., Symeonidis, L., Diamantoglou, S. and S. Karataglis (1993). Plant metal content, growth responses and some photosynthetic measurements on field cultivated wheat growing on ore bodies enriched in Cu. *Physiologia Plantarum,* 88: 307–314.

Laura, O.M., Cecilia, L.M., Jorge, S.P. and P.B. Ines (1992). Effect of lead on onion (*Allium cepa* L.) root growth. *Agricultura Tecnica* (Chile), 52: 312–319.

Peng, Y.K. and Z.J. Guo (1993). The inhibitory effect of lead on the seedling growth and the induced formation of oxidase isozymes in higher plant. *Acta Agriculture Boreali Sinica,* 8: 48–53.

Prasad, K. and T. Devi (1990). Effect of lead on *Phaseolus mungo* and *Phaseolus radiatus. Environ. Eco.,* 8: 534–538.

Saggoo, M.I.S. and Poonam Kumari (1994). *In situ* monitoring of genotoxicity of sewage sludge amended soil by screening natural vegetation. *Poll. Res.,* 13(3): 241–247.

Salim, R., Isa, M., Al Subu, M.M., Sayrafi, S.A. and O. Sayrafi (1995). Effect of irrigation with lead and cadmium on the growth and on the metal uptake of cauliflower, spinach and parsley. *J. Env. Sci. and Hlth.,* 30(A): 831–849.

Saravanan, S., Subramani, A. and A.S. Lakshamanachary (1996). Effect of copper sulphate on seed germination and early seedling growth of green gram (*Vigna radiata* Linn. Wilczek). *Eco. Env. Conserv.,* 2(3&4): 163–164.

Sharma, C.B.S.R. (1982). Plant monitors of Environmental Mutagens. In: *Workshop on Evaluation of Mutagenic and Carcinogenic potentials of Environmental Agents*. Environ. Mutagen Society India, BARC, Bombay, p. 27–39.

Sheoran, I.S. and R. Singh (1996). Influence of cadmium and nickel on photosynthesis and grain yield in wheat. *National J. Pl. Improvement*, 1: 53–59.

Shivanna, K.R. and N.S. Rangaswamy (1992). *Pollen Biology*. Springer-Verlag, Germany.

Subramani, A., Saravanan, S., Tamihiniyan, P. and A.S. Lakshamanchary (1997). Influence of Heavy Metals on Germination and Early Seedling Growth of *Vigna mungo* (L.) Hepper. *Poll. Res.*, 16(1): 29–31.

WHO (1972). Evaluation of certain food additives and the contaminants, *WHO Tech. Rep. Ser.*, pp. 505.

Wong, M.H., Cheung, L.C. and W.C. Wong (1984). Effects of roadside dust on seed germination and root growth of *Brassica chinensis* and *B. parachinensis*. *Sci. Total Env.*, 33(73): 87–102.

Zheng, C.R. and H.M. Chen (1989). Effects of complex pollutants on the growth of rice. *Soils Turang*, 21: 10–15.

Chapter 12

Bioaccumulation of Chromium Trioxide and its Effect on Blood Glucose, Glycogen Content and LDH Activity in the Fish, *Tilapia mossambica*

R. Thangam and A.A. Sivakumar

ABSTRACT

Tilapia mossambica was exposed to sublethal concentrations of, Chromium trioxide, 2,4,6,8 ppm for a period of 24, 48 and 72 hrs and changes in the biochemical contents have been observed. The accumulation of Chromium in musle, liver and kidney was found to be increasing with the increase of exposure period and concentration. The toxicity of heavy metal caused the blood glucose level to decrease with the increase of Chromium trioxide concentration and decrease with the increase of exposure period. The glycogen content in muscle and liver was found to decrease with the increase of concentration and decrease with the increase of exposure period. Increase in the Lactate dehydrogenase activity in the muscle, liver, kidney was observed with the increase in concentration of the heavy metal, Chromium trioxide, and exposure period.

Keywords: *Toxicity, Chromium trioxide, Glucose and glycogen levels, LDH activity, Tilapia mossambica.*

Introduction

There are several cases of Chromium contamination in ground water. The more serious of these cases are due to disposal of electroplating wastes in ponds or settling basins. The toxicity of Chromium to aquatic life is strongly influenced by the form of Chromium and quality of water. The toxicity towards Chromium considerably varies between groups of organisms. Invertebrates and phytoplankton have been found to be more sensitive than fish. Among plants, algae are apparently sensitive to Chromium than higher plants. The toxicity to algae is measured in terms of reduction on growth or in terms of reduction in photosynthesis. Freshwater invertebrates have a wide range of tolerance to Chromium and in general, they have been found to be more sensitive than fish. Experimental studies have suggested that calcium chromate may be that specific carcinogenic agent (Enterline, 1974). However, some investigations have produced cancer in experimental animals with injections of either the trivalent or hexavalent form (Hueper and Payne, 1962).

Materials and Methods

The fingerlings of *Tilapia mossambica* were collected from the Department of Fisheries, Malampuzha in Kerala. Ten healthy fishes were exposed to different concentrations of the chromium trioxide to calculate the LC_{50} value (Finney, 1978). $1/10^{th}$ of 24 hrs LC_{50} value was used for the sub-lethal studies, in which the fishes live for long periods without any visible changes. The fishes were exposed to 24, 48 and 72 hours. Control and treated fishes were sacrificed at the end of the exposure period. Blood and tissue samples (Liver, Kidney, and Muscle) were collected. The blood sample was used to analyse the glucose level. Other tissue samples were used to analyse the chromium level, glycogen, and Lactate dehydrogenase activity. The tissues, weighing about 20mg, were homogenized in tris buffer and centrifuged at 2500 rpm for 15 minutes. The clear supernatant was used for the analysis of all the parameters.

Results and Discussion

In the present study, accumulation of Chromium trioxide in the tissues of the fish, *Tilapia mossambica* was found to be increased, with the increase of concentration of heavy metal. Higher amount of

Chromium was present in the liver (Table 12.1). Van Hoof and Van San (1981) revealed that the rudd (*Scardinium erythrophtalmus*) bioaccumulated Chromium in the gill tissue, followed by the liver. Lowest Chromium accumulation has been observed in the muscle. Similar observations have been made by earlier studies (Nishihara *et al.*, 1985; Lazos *et al.*, 1989; Seymore, 1994; Coetzee, 1996; Wepener, 1997).

Low accumulation of metals in muscles of some clupeiod fishes was reported by Sivakumar *et al.* (1989). The Cadmium accumulation was higher in the order of liver and kidney as reported by Thiruvalluvan *et al.* (1997). Chromium accumulation is highest in the order of liver and kidney. High levels of Cadmium accumulation were reported for liver, kidney of the trout, *Salvelinus fontinalis* by Benoit *et al.* (1976).

Blood glucose level was found to be increasing with the increasing of concentration and decreasing with the increase of exposure period (Table 12.2). Schwarz and Mertz (1959) observed the glucose tolerance factor, which was shown later to contain Chromium, was deficient in animals with impaired glucose tolerance, and that supplemental Chromium improved glucose tolerance. Glucose uptake, glucose use for lipogenesis, glucose oxidation to carbondioxide and glycogenesis increase because of the addition of Chromium plus insulin to animal tissues (Anderson, 1987). Diwan *et al.* (1979), Usha and Ramamurthi (1989) reported the imbalance in hormone release during heavy metal toxicity which probably blocks the absorption of glucose into the tissue with consequential rise in blood glucose levels. Nath and Kumar (1987) and Bhattacharya *et al.* (1987) reported the fall in blood glucose level in fishes due to metal exposure.

Present study indicates the decrease of glycogen content in liver, muscle and kidney with the increase of heavy metal concentration (Table 12.3). Similarly, Banerjee *et al.* (1978) observed the depletion of glycogen content in liver after four weeks of exposure. Sastry and Subhadra (1982) observed the reduction glycogen content in the liver of the catfish, *Heteropnuestes fossilis*. Vijayaram *et al.* (1989) studied the Cadmium induced changes in the decreased glycogen contents in the fish tissues. A similar degree of decrease in glycogen content of Cadmium treated fishes was also reported (Shaffi, 1978; Dubale and Shah 1981; Lowe-Jinde and Niimi, 1984). Many reasons

Table 12.1: Accumulation of Chromium Trioxide in the Muscle, Liver and Kidney of the Fish, *Tilapia mossambica*, Exposed for Different Concentrations and Periods

Exposure Concentration (ppm)	*Muscle*			*Liver*			*Kidney*		
	24 hrs	*48 hrs*	*72 hrs*	*24 hrs*	*48 hrs*	*72 hrs*	*24 hrs*	*48 hrs*	*72 hrs*
2	0.030	0.1015	0.1146	0.060	0.1844	0.2266	0.042	0.1148	0.1398
4	0.068	0.1521	0.1534	0.021	0.2248	0.3128	0.086	0.1700	0.4416
6	0.071	0.1725	0.2029	0.032	0.2805	0.3827	0.096	0.2988	0.6144
8	0.089	0.1817	0.2564	0.037	0.3845	0.4761	0.130	0.3292	0.7354

Values are expressed in percentage.

Each value represents the mean of three replications.

could be attributed to the decrease in tissue glycogen. Shaffi (1978) reported that accumulation of Cadmium in the pancreatic islets of fish would selectively damage the insulin producing cells which in turn have contributed for the observed glycogen.

Table 11.2: Effect of Chromium Trioxide on the Glucose Level in the Blood of *Tilapia mossambica* on Exposed for Different Concentrations and Periods

Exposure Concentration (ppm)	*Exposure Periods*		
	24 hrs	*48 hrs*	*72 hrs*
Control	109	98	85
2	111	101	96
4	125	112	105
6	138	127	110
8	144	138	127

Values are expressed in mg/dl.

Each value represents the mean of three replications.

Table 12.3. Chromium Trioxide on the Glycogen Content in the Muscle, Liver of the Fish, *Tilapia mossambica* Exposed for 24 hrs, 48 hrs and 72 hrs

Exposure Concentration (ppm)	*Muscle*			*Liver*		
	24 hrs	*48 hrs*	*72 hrs*	*24 hrs*	*48 hrs*	*72 hrs*
Control	0.926	0.927	0.925	3.768	3.766	3.764
2	0.766	0.526	0.385	1.309	0.931	0.729
4	0.763	0.484	0.340	1.205	0.867	0.660
6	0.758	0.471	0.295	1.555	0.835	0.596
8	0.694	0.419	0.265	1.091	0.803	0.510

Values are expressed in mg/g.

Each value represents the mean of three replications.

Activity of the Lactate dehydrogenase enzyme was increased in the increase of concentrations of Chromium and increase of exposure period of the fish, in the present study (Table 12.4). Fish favour anaerobic metabolism to meet the energy demands under metal stress as indicated by increased Lactate dehydrogenase activity

Table 12.4: Effect of Chromium Trioxide on the Lactate Dehydrogenase Activity in Kidney of the Fish, *Tilapia mossambica* Exposed for 24 hrs, 48 hrs and 72 hrs

Exposure Concentration (ppm)	*Muscle*			*Liver*			*Kidney*		
	24 hrs	*48 hrs*	*72 hrs*	*24 hrs*	*48 hrs*	*72 hrs*	*24 hrs*	*48 hrs*	*72 hrs*
Control	0.128	0.129	0.128	0.131	0.133	0.129	0.119	0.122	0.121
2	0.149	0.156	0.194	0.172	0.201	0.256	0.132	0.150	0.174
4	0.153	0.166	0.214	0.185	0.205	0.269	0.140	0.151	0.179
6	0.156	0.171	0.217	0.188	0.219	0.277	0.147	0.155	0.204
8	0.161	0.181	0.226	0.196	0.238	0.291	0.159	0.158	0.228

Values are expressed in μ moles of pyruvate liberated min/mg. protein.

Each value represents the mean of three replications.

in muscle as studied by the Vineeta Shukla and Sastry (1998). Similar study showed the increased activity of LDH in the liver of *Gambusia affinis* when treated with nitrate (Archana Kumta and Gaikwad, 1998).

Lactate dehydrogenase, which is a cytoplasmic enzyme, showed marked elevation in activity. LDH is generally associated with cellular metabolic activity. It acts as a pivotal enzyme between the glycolytic pathway and the tricarboxylase acid cycle. Thus non-availability of oxygen inhibition of LDH and simultaneous elevation of LDH suggest a bias towards anaerobic glycolytic pathway. This clearly indicates slower mobilization of glycogen from the liver (Larson 1975; Larson and Haux, 1982).

Lactate is the end product of the glycolytic sequence under anaerobic condition. LDR isozymes have high affinity for pyruvate as the electron acceptor. This accounts for high levels of LDH after treatment to nitrite, causing accumulation of lactate under tissue hypoxia causing their stiffening, affecting transamination. Inanov *et al.* (1977) also observed increase in lactate in the presence of Sodium nitrite.

References

Anderson, R.A. (1987). *Chromium in Trace Elements in Human and Animal Nutrition* Vol.1, 5th Ed., (Ed.) W. Mertz, New York. Academic Press Inc., p. 225–244.

Banerjee, S.K., Dastidar, G. Mukhopadhyay, P.K., and P.V. Dehadri (1978). Toxicity of Cadmium: A comparative study in the airbreathing *Tilapia mossambica. Ind. J. Exp. Biol.*, 1274–1277.

Beneoit, A.D., Leonard, E.N., Christensen, G.M. and J.T. Fiandt (1976). Toxic effects of Cadmium on three generations of Brook trout, *Salvelinus foatinalis. Trans. Amer. Fish. Soci.*, 105(4): 550–560.

Bhattacharya, T., Ray, A.K. and S. Bhattacharya (1987). Blood glucose and hepatic glycogen interrelationship in *Channa punctatus* (*Bloch*). A parameter of nonlethal toxicity bioassay with industrial pollutants. *Indian. J. Exp. Biol.*, 25: 539–541.

Coetzee, L. (1996). Bioaccumulation of metals in selected fish species and the effect of pH on Aluminium toxicity in a cichlid, *Oreochromis mossambicus. M.Sc. Thesis*, Rand Afr. Univ., South Africa.

Diwan, A.D. Hingorani and N. Chandra Sekhram Naidu (1979). Levels of blood glucose and tissue glycogen in two live fishes exposed to industrial effluents. *Bull. Environ. Contam. Toxicol.*, 21:269–272.

Dubale, M.S. and P. Shah (1981). Biochemical alternations induced by Cadmium in the liver of *Channa punctatus. Environ. Res.*, 26: 110–118.

Enterline, P.E. (1974). Respiratory cancer among Chromate workers. *J. Occup. Med.*, 16: 523–26.

Finney, D.J. (1978). *Statistical Methods in Biological Assay*, 3rd edition. Griffin Press, London, pp. 508.

Fruton, J.S. and A. Simmonds (1965). *General Biochemistry*. Asia Publishing House, Chennai.

Hueper, W.C. and W.W. Payne (1962). Experimental studies in metal carcinogensis. Chromium, Nickel, Iron, Arsenic. *Arch. Environ. Health.*, 5: 445–562.

Ivanov, N., Nestrov, N., Profirov, Y., Toucheva, E. and O. Petkava (1977). The effect of Sodium nitrite on the ATP, ADP, AMP, and Lactic acid concentration in the liver, skeletal, heart musculature of rats. *Shivotonovud Nauki.*, 14: 87–91.

Kumta, Archana and S.A. Gaikwad (1998). Effect of Nitrite on succinate dehydrogenase (SDH) and lactose dehydrogenase (LDH) in fresh water fish *Cambusia affinis. Poll. Res.*, 17(2): 177–179.

Larsson, A. (1975). Some biochemical effects of Cadmium of fish In: Sublethal Effects on Toxic Chemicals on Aquatic Animals, (Eds.) Koeman, J.H. and J.J.T.W.A. Strik. Elsevier, Amsterdam, pp. 3–13.

Larsson, A. and C. Haux (1982). Altered carbohydrate metabolism in fish exposed to sublethal levels of Cadmium. *J. Environ. Biol.*, 3: 71–81.

Lazos, E.S., Aggelousis, G. and A. Alexakis (1989). Metal and proximate composition of the edible portion of freshwater fish species. *J. Food Comp. Anal.*, 2: 371–381.

Lowe-Jinde, L. and A.J. Niimi (1984). Short term and long term effects of Cadmium ion glycogen reserves and liver size in rainbow trout *Salmo gairdneri* Richardson. *Arch. Environ. Contam. Toxicol.*, 13: 759–764.

Nath, K. and N. Kumar (1987). Toxic impact of hexavalent Chromium on the blood pyruvate of *Colisa fasciatus. Acta Hydrochi. Hydrobiol.*, 15: 531–534.

Nishihara, T., Shimamoo, T., Wen, K.C. and M. Kondo (1985). Accumulation of Lead, Cadmium and Chromium in several organs and tissues of Carp. *Eisei Kagaku.*, 31(2): 119–123.

Sastry, K.V. and K.M. Subhadra (1982). Chronic effect of Cadmium on some aspects of carbohydrate metabolism in a freshwater catfish, *Heteropneustes fossilis. Water, Air and Soil Poll.*, 20: 293–297.

Schwarz, K. and W. Mertz (1959). Chromium(III) and the glucose tolerance factor. *Arch. Biochem. Biophys.*, 85: 292.

Seymore, T. (1994). Bioaccumulation of metals in *Barbus marequensis* from the olifants River, Kruger National Park and letal levels of Manganese to juvenile *Oreachromis mossambicus. M.Sc. Thesis*, Rand Afr. Univ., South Africa.

Shafi, S.A. (1978). Cadmium intoxication on tissue glycogen content in three freshwater telosts. *Curr. Sci.*, 47: 668–870.

Shukla, Vineeta and K.V. Sastry (1998). Acute toxic effects of Cadmium on various parameters in the freshwater fish. *Proc. Acad. Environ Biol.*, 7(2): 155–160.

Sivakumar, R., Sivakumaran, K.P. and V. Ramaiyan (1989). Heavy metals in some clupeid fishes from Parangipettai waters. *National Symposium on Natural Resources and their Conservation.* Parangipettai.

Thiruvalluvan, M., Nagendran, N. and A. Charles Manaharan (1997). Bioaccumulation of Cadmium and Methyl parathion in *Cyprinus carpio var. communis* (Linn). *J. Env. and Poll.*, 4(3): 221–224.

Usha, R. and R. Ramamurthi (1989). Histological alternations in the liver of teleost *Tilapia mossambica* in response to Cadmium toxicity. *Ecotoxicol. Environ. Safety*, 17(2): 221–226.

Vanhoof, F. and M. Vansan (1981). Analysis of Copper, Zinc, Cadmium and Chromium in fish tissues: A tool for detecting metal caused fish kills. *Chemosphere*, 10(10): 1127–1135.

Vijayram, K., Geraldine, P. Varadarajan, T.S. George John and P. Loganathan (1989). Cadmium induced changes in the biochemistry of an airbreathing fish *Anabas testudinus*. *J. Ecobiol.*, 1(4): 245–251.

Wepener, V. (1997). Metal ecotoxicology of the Olifants River in the Kruger National Park and the effect thereof on fish haematology. *Ph.D. Thesis*. Rand Afr. Univ., South Africa.

Chapter 13

Effect of Sublethal Concentrations of Copper on the Selected Organs of an Estuarine Fish *Mystus gulio* (Hamilton)

S.V.S. Amanulla Hameed, K. Muthukumaravel, A. Subramanian and N. Mayilvakanan

ABSTRACT

Mystus gulio has been exposed to lower (0.35 ppm) and higher (1.05 ppm) concentrations of copper. The observed major changes varied from vacuolation in the epithelial cells, shortened and ruptured lamellae, dilation of blood vessels in the gills and pycnotic nuclei, vacuolation, empty blood vessels to hepatic lesions with necrosis in liver. Other notable features were the shortened and ruptured villi, necrosis of epithelium and vacuolation in the submucosa of the intestine. More severe damages found in the fishes exposed to the higher concentration for 21 days, reflect dose and time dependent toxic nature of copper.

***Keywords**: Copper, Histopathology, Mystus gulio.*

Introduction

Heavy metal contamination, being a major concern of chemical pollution is now attracting wide spread interest. Copper ions are toxic to fish, affecting their behaviour, growth and reproductive capacity (Mount and Stephen, 1969, Mckim and Benoit, 1971; Rice and Harrison, 1978; Gupta *et al.*, 1994; Stien *et al.*, 1997; Gupta, 1999). Copper enters the fish body through nutrients and surrounding water may cause severe histological damages (Magendran, 1990, Rajamanickam 1992, Arellano, 2000). However the study on the effect of copper on the various organs of *Mystus gulio* is meager. Hence the present study aims to probe into the structural alterations manifested in the gill, liver and intestine of *M. gulio* exposed to varying concentrations of copper for 21 days.

Materials and Methods

The fishes were collected from the Agniar estuary near Adirampattinam and acclimatized under laboratory conditions (29 ± 1°C). The LC_{50} for copper for 96 hr was found out by using probit method (Finney, 1971). The 1/10th and 1/3rd of LC_{50} were taken as lower and higher concentrations respectively. Experimental fishes were divided into 3 groups and maintained in the plastic trough containing 20 litre capacity of water. First group was kept at water as control without adding any toxic chemical. The second group and the third group were exposed to lower (0.35 ppm) and higher (1.05 ppm) concentrations of copper for a period of 21 days.

The fish from each group was anaesthetized by chloroform. The gill, liver and portions of intestine were dissected out after 21 days and fixed in 5 per cent formaldehyde. Tissues were processed according to standard histological procedures and paraffin sections of 6μ thickness were taken and stained with haemotoxylin and eosin. Sections were analysed and photomicrographs were taken.

Results and Discussion

The histological nature of normal fish and the remarkable histological changes due to copper exposure in the gill, liver and intestine of *M. gulio* were depicted in the Figures (13.1–13.9). The overall observed results of the present study indicated that more severe histopathological changes have been found in all the tissues of fish under (1/3) 30 per cent sublethal concentration after 21 days exposure. This reflects the time and dose dependent nature of copper toxicant.

Fusion of the adjacent secondary lamellae was found in gills of copper exposed *M. gulio*. Randi *et al.* (1996) reported that this

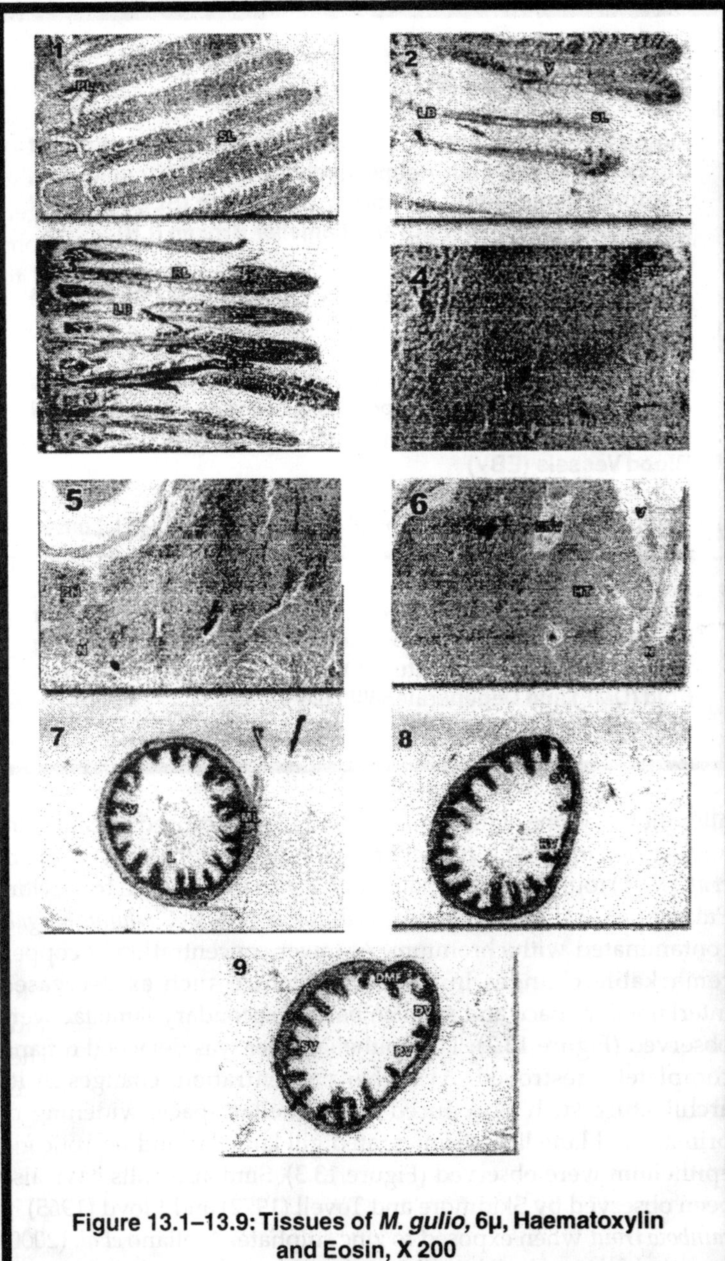

Figure 13.1–13.9: Tissues of *M. gulio*, 6μ, Haematoxylin and Eosin, X 200

Figure 13.1: Gill of Control Fish Showing Primary Lamellae (PL), Secondary Lamellae (SL)

Figure 13.2 & 13.3: Gill of Lower and Higher Concentrations of Copper Exposed Fish Respectively; Showing Vacuoles (V), Ruptured Lamellae (RL), Shortened Lamellae (SL), Fusion of Secondary Lamellae (FSL) and Lamellar Broken (LB)

Figure 13.4: Liver of Control Fish Showing Normal Hepatocytes (NH) and Blood Vessels (BV)

Figure 13.5 & 13.6: Liver of Lower and Higher Concentrations of Copper Exposed Fish Respectively; Showing Pycnotic Nuclei (PN), Necrosis (N), Hypertrophy (HT), Vacuoles (V) and Empty Blood Vessels (EBV)

Figure 13.7: Intestine of Control Fish Showing Villi (V), Lumen (L) and Mucosa Layer (ML)

Figure 13.8 & 13.9: Intestine of Lower and Higher Concentrations of Copper Exposed Fish, Respectively Showing Shortened Villi (SV), Ruptured Villi (RV), Destroyed Mucosal Fold (DMF) and Degenerated Villi (DV)

alteration was one of the branchial pathologies caused by cadmium and clearly implied a reduction in the respiratory surface. Holm *et al.* (1991) found a similar pathology in fish treated with tributyltin. Palanippan *et al.* (2002) observed similar results in *Cirrhinus mrigala* contaminated with chromium. At lower concentration of copper, remarkable changes in the gill structure such as decreased interlamellar space and swollen base of secondary lamellae were observed (Figure 13.2). The epithelial layer was detached off and completely destroyed. At higher concentration, changes in its architecture such as reduced interlamellar space, widening of primary gill lamellae, disintegrated gill lamellae and necrotic gill epithelium were observed (Figure 13.3). Similar results have also been observed by Skidmore and Tovell (1972) and Lloyd (1965) in *rainbow trout* when exposed to zinc sulphate. Arellano *et al.* (2000) reported fusion of gill lamellae and separation of epithelial layer of the secondary lamellae under copper exposure to *Solea senegalensis* and Toad fish *Halobatrachus didactylus.*

Livercord disarray, pycnotic nuclei in hepatic cell, vacuolisation, necrosis, accumulated cytoplasmic granules and empty blood vessels were found in the liver of copper treated *M. gulio* (Figures 13.5–13.6). Toxicants induced changes in the liver of fishes can be recorded as an index of the identification of pollutional stress in fish (Couch, 1975). These observations are in agreement with mercury intoxication reported by Sastry and Gupta (1978). The vacuolation, necrosis and appearance of some typical globular bodies in *Punticus conchonius* due to Zn toxicity have been reported by Kumar and Pant (1981).

Rupture of muscular layer, necrosis, vacuolation, reduction and rupturing of villi were observed in the intestine of *M. gulio* exposed to sublethal concentrations of copper (Figures 13.8–13.9). Similar pathological effects were observed in the intestine of *Sarotherodon mossambicus* exposed to mercury (Akkilender Naidu *et al.*, 1983), *M. vittatus* to chromium (Srivastawa and Maurya, 1991) *Labeo rohita* to mercury chloride (Jagadeesan, 1994) and *Hypothalmichthys molitrix* to nickel and copper (Athikesavan, 2001). Akkilander Naidu *et al.* (1983) suggested that the shortening of intestinal villi was probably due to the dissolution of tissues under toxic conditions. From these results it is clear that the fishes could be used to identify the environmental pollution and stress.

Acknowledgement

The authors wishes to express their sincere thanks to Dr. A. Pannerselvam, Professor, Department of Botany, A.V.V.M Sri Pushpam College, Thanjavur to use Photomicrograph facilities in the laboratory.

References

Akkilender Naidu, K., Abhinender Naidu, K., and R. Ramamurthi (1983). Histopathological alterations in liver and intestine of teleost, *Sarotherodon mossambicus* in response to mercury toxicity. *Ecotoxicol, Environ. Saf, 7(6)*: 566–575.

Arellano, J.M., Blasco, J., Ortiz, J.B., Capeta–Da Silva, D., Navarro, A., Sanchez Del Pino, M.J. and C. Sarasquete (2000).

Accumulation and histopathological effects of copper in gills and liver of *Senegales sole, Solea senegalensis* and Toad fish, *Halobatrachus didactyhis. Ecotoxical. Environ. Res.*, 3(1): 22–28.

Athikesavan, S. (2001). Studies on the impact of heavy metal pollutants (Nickel and zinc) on bioaccumulation, cytogenetics and histological changes of the freshwater fish *Hypophalmichthys molitrix* (valenciennes). *Ph.D. Thesis*, University of Madras, pp. 76.

Couch, J.A. (1975). Histopathological effects of pesticides and related chemicals on the liver of fishes. In: *The Pathology of Fishes*, (Eds.) W.E. Ribelin and G. Migaki. The University of Wisconsin Press, p. 15–26.

Finney, D.J. (1971). *Probit Analysis*, 3rd Ed. Cambridge University Press, Cambridge, pp. 333.

Gupta, A.K. (1999). Short-term toxicity of copper to *Channa punctatus* (Bloach). *J. Environ. Biol.*, 20(3): 227–229.

Gupta, A.K., Sajini, C.B and M.S. Champawat (1994). Acute toxicity of Cu, Cd, Pb and Zn to a freshwater teleost, *Poeciia reticulata* (Peters). *Proc. 3rd Asian Fish Forum*, p. 499–502.

Holm, G.M., Norrgren, I. and O. Linden (1991). Reproductive and histopathological effects of long-term experimental exposure to bis (tributyltin) oxide (TBTO) on the three-spined stickleback. *J. Fish Biol.*, 38: 373–386.

Jagadeesan, G. (1994). Studies on the toxic effects of mercuric chloride and the influence of antidote Dimercaprol on selected tissues in *Labeo rohita* (Hamilton) fingerlings. *Ph.D. Thesis*. Annamalai University, India, pp. 84.

Kumar, S. and S.C. Pant (1981). Histological effects of acute toxic levels of copper and zinc on gills, liver and kidney of *Puncticus conchonius* (Ham). *Ind. J. Exp. Biol.*, 19: 191–194.

Lloyd, R.H. (1965). Factors that effect the tolerance of fish to heavy metal poisoning. Biological problems in the water pollution, 3rd Seminar, U.S. Dept., Health Education and Welfare, pp. 181.

Magendran, A. (1990). Studies of heavy metal (Cu, Cd and Hg) stress on pearlspot, *Etroplus suratensis* (Bloch). *Ph.D. Thesis*. Annamalai University, India, pp. 89.

Mckim, J.M. and D.A. Benoit (1971). Effects of long term exposure to copper on the survival growth and reproduction to brok trout *Salvelinus fontinalis. J. Fish Res. Board Can.*, 28: 655–662.

Mount, D.I. and C.E. Stephen (1969). Chronic toxicity of copper to the fat head minnow (*Pimephales promels*) in soft water. *J. Fish Res. Board Can.*, 26: 2449–2457.

Palaniappan, P.L.R.M., Karthikeyan, S.K. and Selvisabhanayakam (2002). Effect of Sub-lethal concentrations of chromium on gill tissues of an edible fish *Cirrhinus mrigala. J. Curr. Sci.*, 2(2): 271–274.

Rajamanikam, C. (1992). Effect of heavy metal copper on the biochemical contents, bioaccumulation and histology of the selected organs in the freshwater fish, *Mystus vittatus* (Bloch). *Ph.D. Thesis*. Annamalai University, India. Pp. 78.

Randi, A.S., Monserrat, J.M., Rodriguez, E.M. and I.A. Romano (1996). Histopathological effects of cadmium on the gills of the freshwater, *Macropsobrycon uruguayanae.* Eignmann (Pisces, Atherinidae). *J. Fish. Dis.*, 19: 311–322.

Rice, D.W. and F.I. Harrison (1978). Copper sensitivity of Pacific herring *Clupea harengus* pallasi, during its early life history. *U.S. Natl. Mar. Fish. Serv. Bull.*, 76: 347–356.

Sastry, K.V. and P.K. Gupta (1978). Effect of mercuric chloride on the digestive system of *Channa punctatus. Bull. Environ. Contam. Toxical.*, 20: 353.

Skidmore, J.F. and P.W.A. Tovell (1972). Toxic effect of zinc sulphate on the gills of rainbow trout. *Wat. Res.*, 6: 217–230.

Srivastava, V.M.S. and R.S. Maurya (1991). Effect of chromium stress on gill and intestine of *Mystus vittatus* (Bloch): Scanning Electron Microscopic study. *J. Ecobiol.*, 3(1): 59–71.

Stein X., Christine, R., Mauricette, G.B., Michele, R. and L. Marc (1997). Effect of copper chloride *in vitro* and *in vivo* on the hepatic erode activity in the fish *Dicentrarchus labrax. Environ. Toxicol. Chem.*, 16: 214–219.

Chapter 14

Impact of Textile Wastewater on *Raphanus sativus* Var. Pusa Reshmi: A Pot Experiment with Special Emphasis on Analysis of Heavy Metals

Richa Marwari, T.I. Khan and H.S. Sharma

ABSTRACT

Raphanus sativus plant material harvested at pre-flowering, peak-flowering and post-flowering stages in a laboratory experiment using five different levels (Distilled Water : Wastewater) of textile wastewater (Collected from textile industrial area Sanganer, Jaipur) was analyzed for chlorophyll content, protein content, per cent nitrogen content and heavy metals. Root length, shoot length, root dry weight and shoot dry weight were also determined. After the crop harvesting plant material contained 1.198 mg/g dry wt. of Zn, 0.803 mg/g dry wt. of Cu, 0.412 mg/g dry wt. of Ni, 0.211 mg/g dry wt. of Cd, 0.617 mg/g dry wt. of Cr, 0.919 mg/g dry wt. of Pb, 0.207 mg/g dry wt. of Co in the plants treated with the highest ratio of distilled water and wastewater (Distilled Water : Wastewater: 80 : 20). After the crop harvesting the soil was found to contain 1.401 mg/g of Zn, 0.999 mg/g of Cu, 0.562 mg/g of Ni, 0.363 mg/g of Cd, 0.762 mg/g of Cr, 1.124 mg/g of Pb and 0.408 mg/g of Co in the soil of pots treated with highest ratio of distilled water and wastewater.

Keywords: *Textile wastewater, Heavy metals, Raphanus sativus, Human health.*

Introduction

Sanganer town situated 20 km away from the main city of Jaipur, is famous for many thriving cottage industries like dyeing and printing and waste paper recycling, and blue potteries. These industries cover about 12.9 sq.km. urban area of Sanganer. A large number of small, medium and large scale textile industrial units are located in Sanganer. The agony despite the best efforts made, is that there is no common effluent treatment plant installed in Sanganer. The untreated wastewater (from these industries) which contains variety of chemicals (Anelin, Caustic soda, Acids, Bleaching powder etc.) including heavy metals is used in irrigating agricultural fields for growing vegetables and other crop plants. It certainly makes the part of food chain. These vegetables and plants, are consumed by human beings in and around Sanganer in Jaipur district.

Impact of Textile Wastewater on agricultural crop plants has been studied by several workers (Carlson *et al.*, 1975; Berry, 1976; Mukherjee and Maitra, 1976; Mhatre and Chaphekar, 1982; Brown and Wilkins, 1986; Maury *et al.*, 1986; Dayama, 1987; Dubey and Dwivedi, 1987; Kumar, 1989; Agarwal and Agarwal, 1990; Bahadur and Sharma, 1990; Gupta and Nathawat, 1991; Gupta, 1992; Khan and Jain, 1995; Jain and Khan, 1996; Sharma and Rao, 1996; Prasad *et al.*, 1997; Angadi and Mathad, 1998; Shrivastava and Purnima, 1999; Tomar *et al.*, 2000).

In the present study for further investigation a pot experiment is conducted to study the adverse effects of textile wastewater on *Raphanus sativus* Var. Pusa Reshmi, in the laboratory of Indira Gandhi Centre for H.E.E.P.S. Univ. of Rajasthan, Jaipur, India.

Materials and Methods

Seeds for the pot experiments were collected from the Agriculture Research Station Durgapura, Jaipur. The plants were grown in the earthenware pots under the natural conditions after sterilizing them with dilute hypochloride solution. *Raphanus sativus* Var. Pusa Reshmi was selected as the test species. *Raphanus sativus* belongs to family Cruciferae. Radish (*Raphanus sativus*) is very popular root crop, grown throughout India.

Experimental Design of Pot Experiment

Five different levels were maintained by diluting the wastewater with distilled water for conducting pot experiments using *Raphanus sativus*. The levels contained Distilled water and wastewater in the following ratios: 95 : 5, 90 : 10, 85 : 15, and 80 : 20. Initially ten seeds were sown in each pot. After the emergence of plumules a uniform population of 5 plants were maintained in each earthen pot. Three replicates were used for each level. The experiments were terminated after post–flowering stages in these crop plants.

To determine the effect of different wastewater levels, many plant growth parameters were studied, *viz.*, Shoot and root length, dry weight of shoot and root, chlorophyll content of leaves (using Arnon's method 1949), percentage of nitrogen in plant and protein content (using Microkjeldahl's method) phosphorus in plant, carbohydrate estimation (method suggested by Dubious *et al.*, 1951) and heavy metals in plants.

The plants were harvested three times as follows:

1. At pre-flowering stage
2. During peak-flowering stage and
3. At post-flowering stage (physiological maturity).

For dry weight determination, root and shoot were separated and dried in hot air oven at 80°C for 72 hours.

Water Analysis

Samples of water (textile wastewater) used in agricultural fields in Sanganer were collected and analyzed. Some physico-chemical properties like pH, Electrical conductivity, chloride, Total solids, Calcium, Magnesium, and Total Hardness and estimation of metal ions were carried out. From these samples physico-chemical properties were studied according to methods suggested by (APHA, AWWA, WPCF, 1985). This water was utilized for experimental pots.

Heavy Metal Analysis

Atomic Absorption Spectrophotometer (AAS Model 646 Shimadzu-Japan) was used for analysis of heavy metals in soil, water and vegetable samples. Specific hollow cathode lamps were used for different heavy metal.

Results

Root Length

Root lengths were measured 10 cm, 14 cm and 16 cm under the controlled conditions (Distilled Water) at pre-flowering, peak-flowering and post-flowering stages respectively. Root length

Table 14.1: Effects of Textile Wastewater on Root Length, Shoot length, Root Dry Weight, Shoot Dry Weight, Chlorophyll 'a', Chlorophyll 'b', Carbohydrate Concentration, Plant Phosphorus, per cent Nitrogen and Protein Content of *Raphanus sativus* Var. Pusa Reshmi

	D.W.	*95 : 05*	*90 : 10*	*85 : 15*	*80 : 20*
Root Length					
Pre-flowering	10 ± 1.5	8 ± 2.3	7 ± 2.3	6 ± 2.0	5 ± 2.2
Peak-flowering	14 ± 2.4	12 ± 2.2	11 ± 1.5	10 ± 2.7	10 ± 2.3
Post-flowering	16 ± 1.5	14 ± 1.9	13 ± 2.7	12 ± 2.4	11 ± 1.9
Shoot Length (cm)					
Pre-flowering	48 ± 2.5	40 ± 2.4	31 ± 1.9	14 ± 2.2	10 ± 1.9
Peak-flowering	54 ± 3.2	47 ± 2.7	38 ± 2.6	20 ± 2.2	15 ± 2.0
Post-flowering	60 ± 1.2	52 ± 3.1	48 ± 2.0	28 ± 2.7	20 ± 2.6
Root Dry Weight (mg)					
Pre-flowering	41.44 ± 1.2	36.01 ± 3.0	29.08 ± 1.9	22.06 ± 2.4	18.16 ± 2.1
Peak-flowering	54.24 ± 2.8	46.12 ± 1.8	35.16 ± 2.2	26.20 ± 2.5	20.34 ± 2.5
Post-flowering	69.01 ± 2.8	58.01 ± 2.3	46.20 ± 2.5	35.31 ± 2.2	28.04 ± 2.6
Shoot Dry Weight (mg)					
Pre-flowering	18.20 ± 3.4	14.25 ± 2.0	12.25 ± 1.9	11.31 ± 1.5	10.02 ± 2.6
Peak-flowering	28.14 ± 2.3	19.16 ± 1.4	16.29 ± 1.9	14.12 ± 1.9	12.56 ± 2.3
Post-flowering	39.36 ± 1.6	27.13 ± 2.8	18.34 ± 1.9	16.16 ± 2.8	14.21 ± 1.6
Chlorophyll a (mg/g)					
Pre-flowering	5.9612 ± 1.8	5.29291 ± 1.4	5.4652 ± 3.1	4.76763 ± 2.1	4.1594 ± 2.0
Peak-flowering	6.51069 ± 1.9	6.0008 ± 2.0	5.58845 ± 2.0	4.82513 ± 2.3	4.20301 ± 2.5
Post-flowering	6.95682 ± 2.4	6.15302 ± 2.0	5.96852 ± 1.6	5.0334 ± 2.3	4.41128 ± 1.9

Contd...

Table 14.1–Contd...

	D.W.	*95 : 05*	*90 : 10*	*85 : 15*	*80 : 20*
Chlorophyll b (mg/g)					
Pre-flowering	10.07216 ± 2.8	9.2449 ± 1.3	8.50224 ± 2.5	7.5105 ± 2.5	6.79256 ± 2.4
Peak-flowering	11.60046 ± 2.1	9.83848 ± 2.0	8.99907 ± 2.3	8.36566 ± 2.3	7.45566 ± 2.4
Post-flowering	11.93996 ± 1.6	10.92164 ± 3.3	10.11328 ± 2.8	8.66136 ± 1.8	7.75136 ± 2.3
Sugar Concentration (mg/g)					
Pre-flowering	0.0368 ± 0.0024	0.0368 ± 0.0002	0.0344 ± 0.0021	0.024 ± 0.0016	0.0272 ± 0.0002
Peak-flowering	0.0416 ± 0.0001	0.040 ± 0.0019	0.0376 ± 0.0002	0.0352 ± 0.0002	0.0304 ± 0.0001
Post-flowering	0.0456 ± 0.0020	0.0424 ± 0.0001	0.0408 ± 0.0001	0.0368 ± 0.0002	0.0352 ± 0.0017
Plant Phosphorus (mg/g dry wt.)					
Pre-flowering	3.01 ± 1.8	2.99 ± 1.6	2.81 ± 1.9	2.79 ± 2.0	2.52 ± 2.0
Peak-flowering	3.17 ± 2.3	3.03 ± 2.5	2.92 ± 1.6	2.83 ± 1.8	2.64 ± 2.2
Post-flowering	3.29 ± 2.0	3.09 ± 1.8	3.03 ± 2.3	2.87 ± 1.3	2.72 ± 2.4
% Nitrogen					
Pre-flowering	0.792 ± 0.15	0.672 ± 0.12	0.659 ± 0.11	0.562 ± 0.16	0.502 ± 0.13
Peak-flowering	1.429 ± 1.6	1.394 ± 1.5	1.270 ± 1.2	1.188 ± 1.5	1.037 ± 1.3
Post-flowering	2.794 ± 1.4	2.686 ± 1.2	2.658 ± 1.6	2.548 ± 1.6	2.085 ± 1.2
Protein (mg/g)					
Pre-flowering	4.95 ± 2.3	4.2 ± 2.4	4.118 ± 2.0	3.512 ± 2.4	3.137 ± 2.3
Peak-flowering	8.931 ± 1.8	8.712 ± 1.9	7.937 ± 2.3	7.425 ± 2.6	6.481 ± 2.7
Post-flowering	17.462 ± 2.0	16.787 ± 2.6	16.612 ± 2.4	15.925 ± 3.1	13.031 ± 3.0

reduced 50 per cent, 28.58 per cent and 31.25 per cent at 80 : 20 (Distilled Water : Wastewater) at the pre-flowering, peak-flowering and post-flowering stages respectively (Table 14.1).

Shoot Length

Shoot length were measured 48 cm, 54 cm and 60 cm under the controlled conditions (Distilled Water) at pre-flowering, peak-flowering and post-flowering stages respectively. Shoot lengths reduced 79.17 per cent, 72.23 per cent and 66.67 per cent at 80 : 20 (Distilled Water : Wastewater) at the pre-flowering, peak-flowering and post-flowering stages respectively (Table 14.1)

Root Dry Weight

Root dry weight was observed 41.44 mg, 54.24 mg and 69.01 mg under the controlled conditions (Distilled Water) at pre-flowering, peak-flowering and post-flowering stages respectively. Root dry weight reduced 56.18 per cent, 62.5 per cent and 59.37 per cent at 80 : 20 (Distilled Water : Wastewater) at the pre-flowering, peak-flowering and post-flowering stages respectively (Table 14.1).

According to the observations it seems that there was more adverse effect on the root dry weight at peak-flowering stage than the pre-flowering stage and post-flowering stage.

Shoot Dry Weight

Shoot dry weight was observed 18.20 mg. 28.14 mg and 39.36 mg under the controlled conditions (Distilled Water) at pre-flowering, peak-flowering and post-flowering stages respectively. Shoot dry weight reduced to 44.95 per cent, 55.37 per cent and 63.9 per cent at 80 : 20 (Distilled Water : Wastewater) at the pre-flowering, peak-flowering and post-flowering stages respectively (Table 14.1).

Chlorophyll Content

Chlorophyll was estimated in milligram per gram (mg/g) dry weight of leaves.

Chlorophyll-a

Chlorophyll-a was estimated 5.9612 mg/g, 6.51069 mg/g and 6.95682 mg/g under the controlled conditions (Distilled Water) at pre-flowering, peak-flowering and post-flowering stages respectively. Chlorophyll-a decreased to 30.23 per cent, 35.45 per cent and 36.6 per cent at 80 : 20 (Distilled Water : Wastewater) at the pre-flowering, peak-flowering and post-flowering stages respectively (Table 14.1).

Chlorophyll-b

Chlorophyll-b was estimated 10.071216 mg/g, 11.60046 mg/g and 11.93996 mg/g under the controlled conditions (Distilled Water) at pre-flowering, peak-flowering and post-flowering stages respectively. Chlorophyll-b decreased to 32.57 per cent, 35.73 per cent and 35.09 per cent at 80 : 20 (Distilled Water : Wastewater) at the pre-flowering, peak-flowering and post-flowering stages respectively (Table 14.1).

Carbohydrates (Soluble Sugar)

Carbohydrates were estimated 0.0368 mg/g, 0.0416 mg/g and 0.0456 mg/g under the controlled conditions (Distilled Water) at pre-flowering, peak-flowering and post-flowering stages respectively. Carbohydrates decreased to 26.09 per cent, 26.93 per cent and 22.81 per cent at 80 : 20 (Distilled Water : Wastewater) at the pre-flowering, peak-flowering and post-flowering stages respectively (Table 14.1).

Plant Phosphorus

Plant phosphorus was estimated 3.01 mg/g, 3.17 mg/g and 3.29 mg/g under the controlled conditions (Distilled Water) at pre-flowering, peak-flowering and post-flowering stages respectively. Plant phosphorus decreased to 16.28 per cent, 16.72 per cent and 17.33 per cent at 80 : 20 (Distilled Water : Wastewater) at the pre-flowering, peak-flowering and post-flowering stages respectively (Table 14.1).

Nitrogen Content

Nitrogen content was estimated 0.792 mg/g, 1.429 mg/g and 2.794 mg/g under the controlled conditions (Distilled Water) at pre-flowering, peak-flowering and post-flowering stages respectively. Nitrogen content decreased to 36.62 per cent, 27.44 per cent and 25.38 per cent at 80 : 20 (Distilled Water : Wastewater) at the pre-flowering, peak-flowering and post-flowering stages respectively (Table 14.1).

Proteins

Proteins were estimated 4.95 mg/g, 8.931 mg/g and 17.462 mg/g under the controlled conditions (Distilled Water) at pre-flowering, peak-flowering and post-flowering stages respectively. Protein decreased to 36.63 per cent, 27.44 per cent and 25.38 per cent at 80 : 20 (Distilled Water : Wastewater) at the pre-flowering, peak-flowering and post-flowering stages respectively (Table 14.1).

Heavy Metal Analysis of Composite Samples of *Raphanus sativus* Var. Pusa Reshmi from Fot Culture (at Flowering Stage)

Zn

The Zn concentration was 0.042 under controlled condition (Distilled Water). The Zn concentration increased as the textile wastewater treatment level increased. At wastewater treatment level 80 : 20, the Zn was observed 1.154 mg/g dry wt, 1.171 mg/g dry wt and 1.198 mg/g dry wt. at pre-flowering, peak-flowering and post-flowering stages respectively.

Cu

The Cu concentration was 0.06.1 under controlled condition (Distilled Water). The Cu concentration increased as the textile wastewater treatment level increased. At wastewater treatment level 80 : 20, Cu was observed 0.687 mg/g dry wt, 0.715 mg/g dry wt and 0.803 mg/g dry wt at pre-flowering, peak-flowering and post-flowering stages respectively.

Ni

The Ni concentration was 0.040 under controlled condition (Distilled Water). The Ni concentration increased as the textile wastewater treatment level increased. At wastewater treatment level 80 : 20, the Ni was observed 0.387 mg/g dry wt, 0.398 mg/g dry wt and 0.412 mg/g dry wt at pre-flowering, peak-flowering and post-flowering stages respectively.

Cd

The Cd concentration was 0.005 under controlled condition (Distilled Water). The Cd concentration increased as the textile wastewater treatment level increased. At wastewater treatment level 80 : 20, the Cd concentration was observed 0.185 mg/g dry wt, 0.198 mg/g dry wt and 0.211 mg/g dry wt at pre-flowering, peak-flowering and post-flowering stages respectively.

Table 14.2: Heavy Metal Analysis of *Raphanus sativus* Var. Pusa Reshmi in Pot Experiment at Pre, Peak and Post-flowering Stages

Stages	*Levels*	*Heavy Metal mg/g' d. wt.*							*Total Heavy Metals*
		Zn	*Cu*	*Ni*	*Cd*	*Cr*	*Pb*	*Co*	
Pre	D.W.	0.042	0.061	0.040	0.005	0.060	0.042	0.040	0.290
Flowering	95 : 05	0.432	0.178	0.082	0.011	0.023	0.112	0.065	0.903
Stage	90 : 10	0.808	0.279	0.100	0.029	0.112	0.301	0.089	1.718
	85 : 15	1.055	0.488	0.185	0.077	0.332	0.511	0.098	2.746
	80 : 20	1.154	0.687	0.387	0.185	0.511	0.881	0.112	3.917
Peak	D.W.	0.042	0.061	0.040	0.005	0.060	0.042	0.040	0.290
Flowering	95 : 05	0.501	0.197	0.082	0.039	0.071	0.191	0.077	1.158
Stage	90 : 10	0.818	0.311	0.097	0.057	0.198	0.344	0.092	1.917
	85 : 15	1.076	0.514	0.199	0.097	0.397	0.577	0.108	2.968
	80 : 20	1.171	0.715	0.398	0.198	0.588	0.911	0.191	4.172
Post	D.W.	0.042	0.061	0.040	0.005	0.060	0.042	0.040	0.290
Flowering	95 : 05	0.515	0.212	0.097	0.048	0.089	0.201	0.065	1.227
Stage	90 : 10	0.888	0.329	0.110	0.069	0.207	0.395	0.080	2.078
	85 : 15	1.097	0.565	0.202	0.105	0.415	0.612	0.105	3.101
	80 : 20	1.198	0.803	0.412	0.211	0.617	0.919	0.207	4.367

Cr

The Cr concentration was 0.060 under controlled condition (Distilled Water). The Cr concentration increased as the textile wastewater treatment level increased. At wastewater treatment level 80 : 20 Cr was observed 0.511 mg/g. dry wt, 0.588 mg/g dry wt and 0.617 mg/g dry wt at pre-flowering, peak-flowering and post-flowering stages respectively.

Pb

The Pb concentration was 0.042 under controlled condition (Distilled Water). The Pb concentration increased as the textile wastewater treatment level increased. At wastewater treatment level 80 : 20, Pb was observed 0.881 mg/g dry wt, 0.911 mg/g dry wt and 0.919 mg/g dry wt at pre-flowering, peak-flowering and post-flowering stages respectively.

Co

The Co concentration was 0.040 under controlled condition (Distilled Water). The Co concentration increased as the textile wastewater treatment level increased. At wastewater treatment level 80 : 20, the Co was observed 0.112 mg/g dry wt, 0.191 mg/g dry wt and 0.207 mg/g dry wt at pre-flowering, peak-flowering and post-flowering stages respectively.

Table 14.3: Heavy Metal Analysis of Pot Soil After Harvesting of *Raphanus sativus* Var. Pusa Reshmi

Level	*Heavy Metals mg/g*							*Total Heavy Metals*
	Zn	*Cu*	*Ni*	*Cd*	*Cr*	*Pb*	*Co*	
D.W.	0.020	0.040	0.020	0.020	0.020	0.020	0.020	0.160
95 : 05	0.231	0.101	0.063	0.052	0.092	0.101	0.067	0.707
90 : 10	0.419	0.201	0.116	0.102	0.173	0.263	0.105	1.379
85 : 15	0.718	0.502	0.284	0.201	0.384	0.575	0.216	2.88
80 : 20	1.401	0.999	0.562	0.363	0.762	1.124	0.408	5.619

Discussions

Khan *et al.* (1997) analyzed five crop plants *viz.*, Fenugreek (*Trigonella foenum graecum*), cauliflower (*Brassica olereacea*), pigweed

(*Chenopodium album*), castor (*Ricinus communis*) and mustard (*Brassica campestris*). Chlorophyll a,b, protein, carbohydrate and heavy metal contents were estimated. The concentrations of Cu, Fe, Pb, Zn, Cd, Cr, Hg and Ni were found to be very high in the vegetables. These metals are consumed by human beings through vegetables. Concentrations of heavy metals were estimated in soil, plants (*Raphanus sativus*), effluent water in the present study.

Reduction in growth of plants was observed in the present study due to textile wastewater. Bhatt and Gokhale (1984) have discussed the various aspects of pollution control problems of the textile, dyestuff and dye intermediates and processing industries. The corrective measures that can be implemented by these industries and techno-economics of effluent treatment plants have also been discussed.

In the present study the effect of textile wastewater (with different levels) has been observed experimentally. It was observed that plant growth in the pots containing higher concentration of textile waste was adversely affected. Significantly, Root length, Root dry weight, Shoot length, Shoot dry weight, Chlorophyll content, Carbohydrate content, Plant phosphorus, per cent Nitrogen and Protein content were the parameters effected adversely.

Wang and Waygood (1959) observed that nickel inhibited the synthesis of chlorophyll in detached leaves of wheat. Mudri *et al.* (1970) studied the treatment of wastes from a dye and printing industry and suggested suitable treatment methods which include removal of zinc by pH adjustment to 9.2 and colour by chlorination. Cadmium at low concentration is not toxic to plants, although in some species it may get accumulated in leaves which are toxic to animals in the food chain (Vallee and Ulmer, 1972) but at higher concentration it is toxic to plants and cause chlorosis in leaves accompained by inhibition of photosynthesis (Huang *et al.*, 1974). Dolor *et al.* (1972) reported that the reduction in plant system is a complex phenomenon resulting from the combined toxic effects of excessive quantity of micronutrient, heavy metals, decomposition products and salt toxicity as well as effects on soil texture and aeration. Numerous studies have shown a significant reduction of growth with increasing lead concentration. These studies were carried out on rice and oat (Fiussello and Molinari, 1973), Soybean (Huang *et al.*, 1974), corn (Carlson *et al.*, 1975). Huang *et al.* (1974)

also reported that lead inhibited pod fresh weight in soybean by 35 per cent in comparison to the control. Mukherji and Maitra (1976) observed that root growth was completely inhibited and germination ceased altogether in *Avena sativa*. Fresh weight of seedling decreased by about 34 per cent and dry weight by about 23 per cent in comparison to control, when treated with mercury compounds. Mukherji and Mukherji (1979) studied the effect of lead on rice. They found complete inhibition of root growth and germination of rice at 2×10^{-2} M and 10^{-1} M lead concentration respectively. Inhibition in root growth was stronger than shoot growth. It appears that the root tip meristem in direct contact with the metals caused such retardation in the growth of roots. Heavy metals bring a considerable decrease in the rate of all the metabolic processes of cells including nucleic acid synthesis, cell division and protein contents. (Nag *et al.*, 1981).

Dubey and Dwivedi (1987), Ray and Khangarot (1989), Agarwal and Agarwal (1990), Bahadur and Sharma (1990), Kumar *et al.* (1991) and Sharma and Rao (1996) have studied the effects of textile wastewater on germination behavior of various crop plants.

Conclusion

Besides the adverse impact of the textile wastewater on growth parameters there is yet another dimension in this study. The amount of heavy metals (Zinc, Copper, Nickel, Cadmium, Chromium, Lead, Cobalt) accumulated in the roots remains the matter of concern, as the root part of Raphanus is consumed by the people of Jaipur. Heavy metals are the known carcinogenic agents. *Raphanus sativus* is widely used a vegetable and eaten up freshly also by the people of all socio-economic groups. Thus putting large number of people at high risk of cancers of different nature.

Hence, there is a need of immediately stopping the practice of growing vegetables in Sanganer with textile wastewater.

Acknowledgement

Authors are thankful to Director, Indira Gandhi Center for H.E.E.P.S., University of Rajasthan for facilities, and encouragement.

References

Agarwal, R. and S.K. Agarwal (1990). Physico-chemical characteristics of Kota saree printing effluent, its effect on seed germination and seedling growth of *Cyamopsis tetragonoloba. Acta Ecol.*, 12(2): 112–118.

Angadi, S.B. and P. Mathad (1998). Effect of copper, cadmium and mercury on the morphological, physiological and biochemical characteristics of *Scenedesmus quadricauda* (Turp) de Breb. *J. Environ. Biol.*, 19(2), 119–124.

Arnon, D.I. (1949). Copper enzymes in isolated chloroplasts. Polyphenoloxidase in *Beta vulgaris. Plant Physiol.*, 24: 1–15.

Bahadur, B. and B.K. Sharma (1990). Effect of industrial effluent on seed germination and early seedling growth of *Tritium aestivum* var. UP 115. *Acta Botanica Indica*, 18: 80–83.

Berry, W.L. (1976). The effects of Zinc on the dose response cause of seedling lettuce to acute Ni-toxicity. *Agron. Abst.*, pp. 20.

Bhatt, S.R. and S.V. Gokhale (1984). Pollution control in textile and allied industries of India. *Mazingira*, p. 1–40.

Brown, M.T. and D.A. Wilkins (1986). The effect of zinc on germination, survival and growth of Betula seeds. *Environ. Pollut.* Ser. A., 41(1): 53–61.

Carlson, R.W., Bazaaz F.A. and G.L. Rolfe (1975). The effect of heavy metals on plants, II. Net photosynthesis and transpiration of whole corn and sunflower plants treated with Pb, Cd, Ni. *Environ. Res.*, 10: 113–120.

Dayama, O.P. (1987). Influence of dyeing and textile water pollution on nodulation and germination of gram. *Act. Ecol.*, 9(1,2): 34–37.

Dolor, S.D., Boyle, J.R. and A.D. Kenney (1972). Paper mill sludge disposal on soils: Effects on the yield and mineral nutrition of oats (*Avena Sativa*). *J. Env. Qual.*, 1: 406–409.

Dubey, R.C. and R.S. Dwivedi (1987). Effect of heavy metals on seed germination and seedling growth of soyabean. *Nat. Acad. Sci. Lett.*, 10(4): 121–123.

Dubious, M., Gilles, K., Hamilton, J.K., Rebers, P.A. and F. Smith (1951). A colorimetric method for the determination of sugar. *Nature*, 16: 167–169.

Fiussello, N. and M.T. Molinari (1973). Effects of lead on plant growth. *Allionia*, 19: 89–96.

Gupta, A.K. and G.S. Nathawat (1991). Effect of textile effluent on germination and seedling growth of *Pisum sativum* var. RPG–3. *Act Ecol.*, 13(2): 109–112.

Gupta, A.K. (1992). A study of effluents of textile and paper industries in Sanganer town and their effects. *Ph.D. Thesis*. University of Rajasthan, Jaipur.

Huang, C.Y., Bazzaz, F.A. and L.N. Vanerhoff (1974). The inhibition of soyabean metabolism by cadmium and lead. *Plant Physiol.*, 54: 122–124.

Jain, V. and T.I. Khan (1996). Effects of wastewater from the textile industry on *Cyamopsis tetragonoloba* var. RGC. 1986. *Int. J. Env. Edu. & Inf.*, 15(1): 67–72.

Khan, T.I., Jain, V. (1995. Effects of textile industry wastewater on growth and some biochemical parameters of *Triticum aestivum* var. Raj 3077. *Jour. Env. and Poll.* 2(2),47–50.

Khan, T.I., Kaur, N., Agrawal, M. (1997). Heavy metal analysis of crop plants from Agricultural fields of Sanganer town. *J. Env. & Poll.*, 4(1): 35–38.

Kumar, P., Patel, B. and K.T.R. Kumar (1991). Effect of Pharmaceutical factory effluent on seed germination, seedling growth and pod productivity of *Mustard.Geobios*, 18: 91–97.

Kumar, S. (1989). Effect of seed treatment with Ni, Cd and Zn on seedling growth of two cultivars of *Hordeum vulgare. Geobios*, 16: 15–20.

Maury, A.N., Gupta R.K. and A.K. Gupta (1986). Effect of heavy metal pollution on seed germination of four Pasture Plants of U.P. *Indian J. Env. Hlth.*, 29(2): 134–139.

Mhatre, G.N. and S.B. Chaphekar (1982). Effect of heavy metals on seed germination and early growth. *J. Environ. Biol.*, 3(2): 53–63.

Mudri, S.S., Kulkarni, A.L., Subrahmanyam, P.V.R. and G.J. Mohanrao (1970). Treatment of wastes from a dye factory. *Env. Hlth.*, 12: 201–217.

Mukherji, C. and S. Mukherji (1979). Effect of mercury on the induction of thermo sensitivity of germinating rice seeds and permeability of potato tuber and beet root tissue. *Indian J. Exp. Biol.*, 15: 422–424.

Mukherji, S. and P. Maitra (1976). Toxic effects of lead on growth and metabolism of germinating rice (*Oryza sativa* L.) seeds as influenced by toxic concentration of lead. Z. *Pflanzenphysiol.*, 81: 26–33.

Nag, P., Paul, A.K. and S.K. Mukherji (1981). Heavy metal effects in plant tissues involving Chlorophyll, Chlorophylase, Hill reaction activity and Gel Electrophoratic patterns of soluble proteins *Indian J. Exp. Biol.*, 19: 702–706.

Prasad, S.D., Rana, T., Singh, P. and K.V. Sastry (1997). Accumulation of chromium and nickel in wheat in a field irrigated with industrial effluents and water hyacinth in Sonipat city, Haryana, India. *J. Env. Biol.*, 18(1): 33–36.

Ray, P.K. and B.S. Khangarot (1989). The Deadly metals. *Science Reporters*, 12: 560–561.

Sharma, A. and C.L.S.N. Rao (1996). Effect of gelatin factory effluents on seed germination and seedling growth of some important crop plants. *J. Env. Biol.*, 17(2): 143–148.

Shrivastava, A.K. and Purnima (1999). Some facets of heavy metal tolerance in higher plants. *J. Indian Bot. Soc.*, 78: 271–286.

Tomar, M., Kaur, I., Neelu and A.K. Bhatnagar (2000). Effect of enhanced lead in soil on growth and development of *Vigna radiata* (Linn.) wilczek. *Indian J. Plant Physiol.*, 5(1)(NS): 13–18.

Vallee, B.L. and D.D. Ulmer (1972). Biochemical effects of Hg, CD and Pb. *Ann. Rev. Biochem.*, 41: 91–128.

Wang, D. and E.F. Waygood (1959). Effect of Benzimidazole and Ni on the Chlorophyll metabolism of detached leaves of Khapti Wheat. *Can. J Bot.*, 37: 743–749.

Chapter 15

Bioaccumulation of Metals in Different Food Fishes in Wastewater Fed Wetlands

Paulomi Maiti and Samir Banerjee

ABSTRACT

Sewage fed ponds are widely used for fish production, where recycling of organic loads occur to provide nutrient to the growing fish. However metallic toxicants present in such system may pose several health hazards to those who depend on such fish for food.

Bioaccumulation of metals may vary in the different tissues of the various fish species, according to their nutritional level and body weight. In the present study, quantitative estimation of Cu, Pb, Zn and Cd were carried out in three fish species inhabiting three different trophic strata of the pond ecosystem. Cu increases in liver with age, in all the three fish species. But compared to Labeo rohita (61.68 mg/g) and Catla catla (20.69 mg/g) liver of Oreochromis nilotica accumulates the most (90 mg/g). The metal in the muscle and kidney remains at low and constant level. Zn level also increases with chronic exposure in all the three tissues where liver is the main site for accumulation, Pb and Cd shows no significant variation with age.

***Keywords**: Cu, Cd, Pb, Zn, Sewage, Fish, Bioaccumulation, Water pollution.*

Introduction

Metal bioaccumulation has been one of the public health concerns in waste-fed aquaculture, especially in polyculture systems in which various species of fish are stocked to fully utilize the available niche of the environment and all the available energy resources derived from the waste materials (Lliang *et al.*, 1998).

Presently a number of sewage-fed ponds are located in the eastern part of Kolkata, West Bengal. The area covers about 25,000 hectares from which approximately 7000 tonnes of fish are produced annually which cater about 20 per cent of the city's daily requirements (Gupta and Mitra, 1997). These wetlands receive city sewage and effluents from about 11,000 industries, contaminated with significant amount of metallic ions.

Metal levels in fish increases with the increment of the metal levels in water, sediment and fish food organisms. Bioaccumulation and toxicity of metals in fish is generally influenced by season, temperature, hardness, salinity and pH of water. Microhabitat utilization, feeding habit, age, sex and species of fish also determine the accumulation pattern of metals. Metallic interaction, presence of organic and inorganic ligands in the aquatic system is further important for bioaccumulation of metals in fish (Kotze, 1999; Giessing, 1981; Newman and McIntosh, 1983; Graney *et al.*, 1984; Unlu and Fowler, 1997). Moreover body size which is closely related to fish growth and metabolism has been shown to attribute most of the variation in metal content of fish (Moriarty *et al.*, 1984).

Metals may enter the body of fish by gills, food, or by diffusion through body surface. Food and drinking water is the major route of uptake of metals in fish, with gills and skin acting as both for uptake as well as excretory route. Kidney and intestine acts as the main route of metal excretion. Liver and kidney act for storage and biotransformation of xenobiotics (Heath, 1995). Thus the accumulation of metals by various intracellular ligand pools is determined by the relative rates of metal binding and release. The rate of metal binding for a given ligand pool is determined by the relative metal binding affinity and size of the pool (Roesjadi and Klerks, 1989) and competition among different metals for these sites (Hamilton and Mehrle, 1987).

The present study aims to assess the pattern of accumulation of Cu, Pb, Zn and Cd in three different fish species widely used in sewage-fed culture system, differing in food habit and inhabiting the different trophic strata of the pond ecosystem. *Labeo rohita,* is predominantly a herbivorous column feeder, *Catla catla,* a zooplankton phagic surface feeder while *Oreochromis nilotica,* feeds on both vegetable and animal matters. They are reared together in waste-fed wetlands. Two different weights of the fishes (50 g and 1000 g) are considered to access the pattern of change in metal accumulation related to their growth and body size.

Site of Study

The study was carried out in a sewage-fed farm "The East Calcutta Fisherman's Co-operative Society" situated at the Eastern Metropolitan Bypass near the Kasba connector (22°30′ N, 88°25′ E). The farms covers an area of 44.55 hectares of land and supporting about 109 sewage fed ponds. Variable quantity of city sewage is frequently discharged into these shallow water bodies regularly throughout the season. A huge organic load is recycled in these ponds as nutrients to nourish a number of culturable food fish for low cost production of animal protein.

Materials and Methods

For the present investigation, *Labeo rohita, Catla catla* and *Oreochromis nilotica* of 2 different body weights (50g and 1000g) were collected from these ponds and the tissues were analysed for the quantitative accumulation of the metals in them. Fishes were dissected out and the different tissues like muscle, liver and kidney were properly processed for analysis of Cu, Pb, Zn and Cd. Tissues were wiped dry and weighed accurately to 1 g each for acid digestion, following to the procedure of Chernoff (1975). Samples of fish tissues were taken in Borosil hard glass test tubes. For each gram of the sample, 5ml of conc. HNO_3 acid was added and then digested overnight at room temperature. The mixture was placed in a hot plate at 85 ± 5°C and 5 ml (3 : 2 conc. Sulfuric acid : Perchloric acid) was then added to it. The digestion was carried out until the mixture turned into a transparent solution. The mixture was cooled and filtered through an acid soaked filter paper and was adjusted to the required volume with distilled water. Metals were detected in atomic absorption spectrophotometer (Varian AA.575).

The actual concentration was calculated on the basis of total amount of the sample taken and expressed in µg/g. Comparison between metal concentration in the tissues of the fish with various body weights were performed by one way analysis of variance (ANOVA).

Results

Tables 15.1–15.4, show the metal distribution in the fish species in the sewage-fed wetlands of Kolkata.

Copper (Table 15.1; Figures 15.1–15.3)

Copper accumulation is more in liver than that in either muscle or kidney. Fish with 1000 g of body's weight show significant increase in Cu content in all the three species under study, (61.68 µg/g, 20.69 µg/g and 90.00 µg/g in *Labeo rohita, Catla catla* and *Oreochromis nilotica* respectively). The latter accumulating the most muscle shows a low accumulation of the metal (2.25 µg/g, 3.64 µg/g in 50 g and 1000 g of *Labeo rohita* respectively and 2.18 µg/g and 3.59 µg/g in 50g and 1000g of *Catla catla*, respectively). The increase in metal content in the tissue with age is not significant. However decrease in the Cu concentration in the muscle of *Oreochromis nilotica* (5.58 µg/g and 3.90 µg/g in 50 g and 1000 g of fish respectively) possibly indicate the removal of the metal either by liver or by other organs. Low Cu content in kidney of all the three species, signify that the organ does not play a significant role in metal excretion in fish.

Table 15.1: Accumulation of Copper in Fish Tissues

Sl.No.	*Tissues*	*Body Weight of Fish (g)*	
		50	*1000*
1.	*Labeo rohita*		
	Muscle	2.25 ± 0.61	3.64 ± 1.00
	Liver	12.70 ± 1.97	61.68 ± 17.74
	Kidney	4.38 ± 1.73	5.66 ± 1.50
2.	*Catla catla*		
	Muscle	2.18 ± 0.78	3.59 ± 0.80
	Liver	5.92 ± 2.24	20.69 ± 6.80
	Kidney	2.20 ± 0.71	4.21 ± 0.56
3.	*Oreochromis nilotica*		
	Muscle	5.85 + 1.02	3.90 ± 0.42
	Liver	20.20 ± 0.99	90.00 ± 10.20
	Kidney	2.92 ± 0.80	5.60 ± 1.04

Mean ± SD of five different observation; $P < 0.05$.

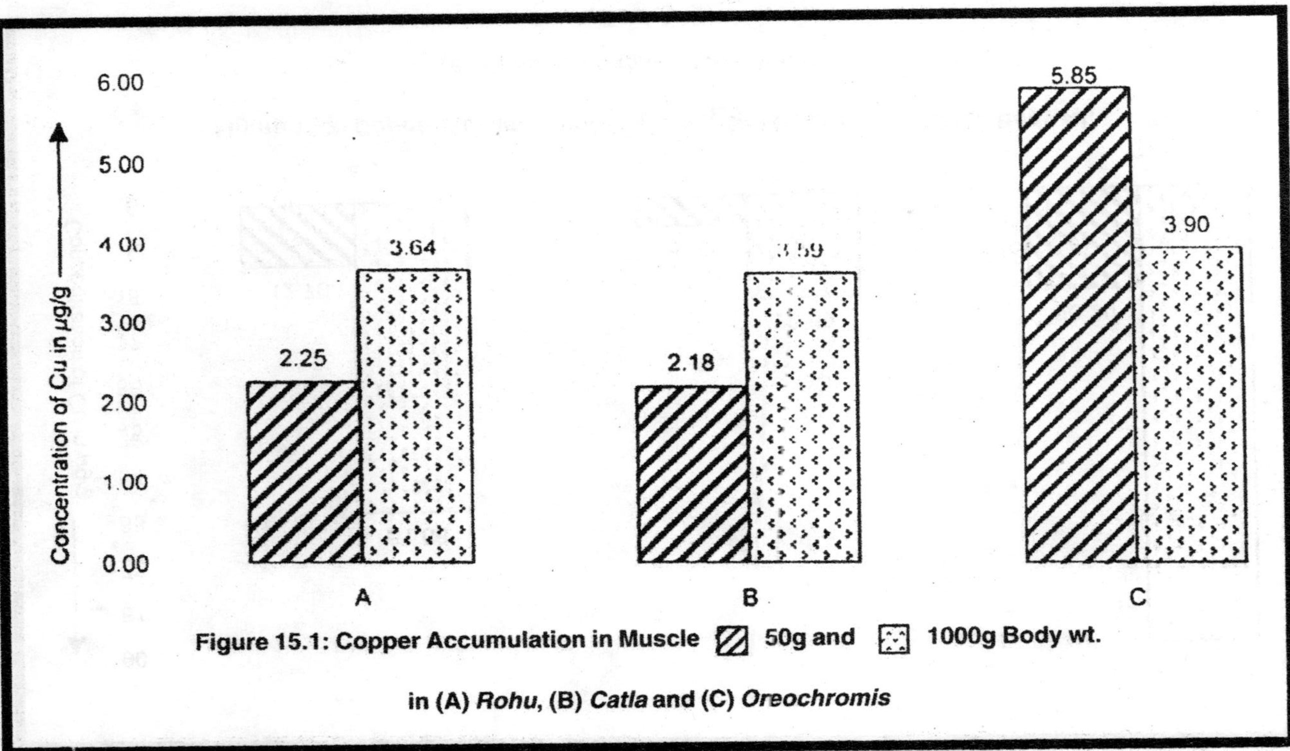

Figure 15.1: Copper Accumulation in Muscle 50g and 1000g Body wt. in (A) *Rohu*, (B) *Catla* and (C) *Oreochromis*

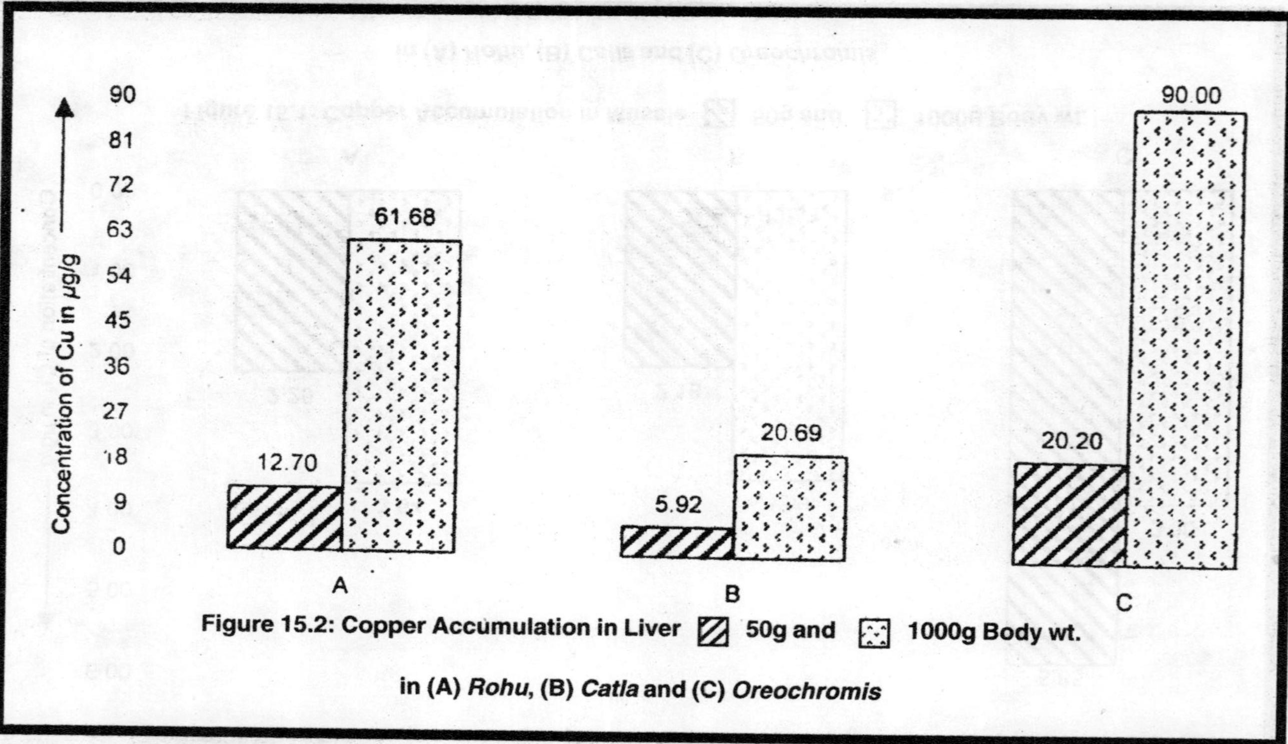

Figure 15.2: Copper Accumulation in Liver (hatched) 50g and (dotted) 1000g Body wt. in (A) *Rohu*, (B) *Catla* and (C) *Oreochromis*

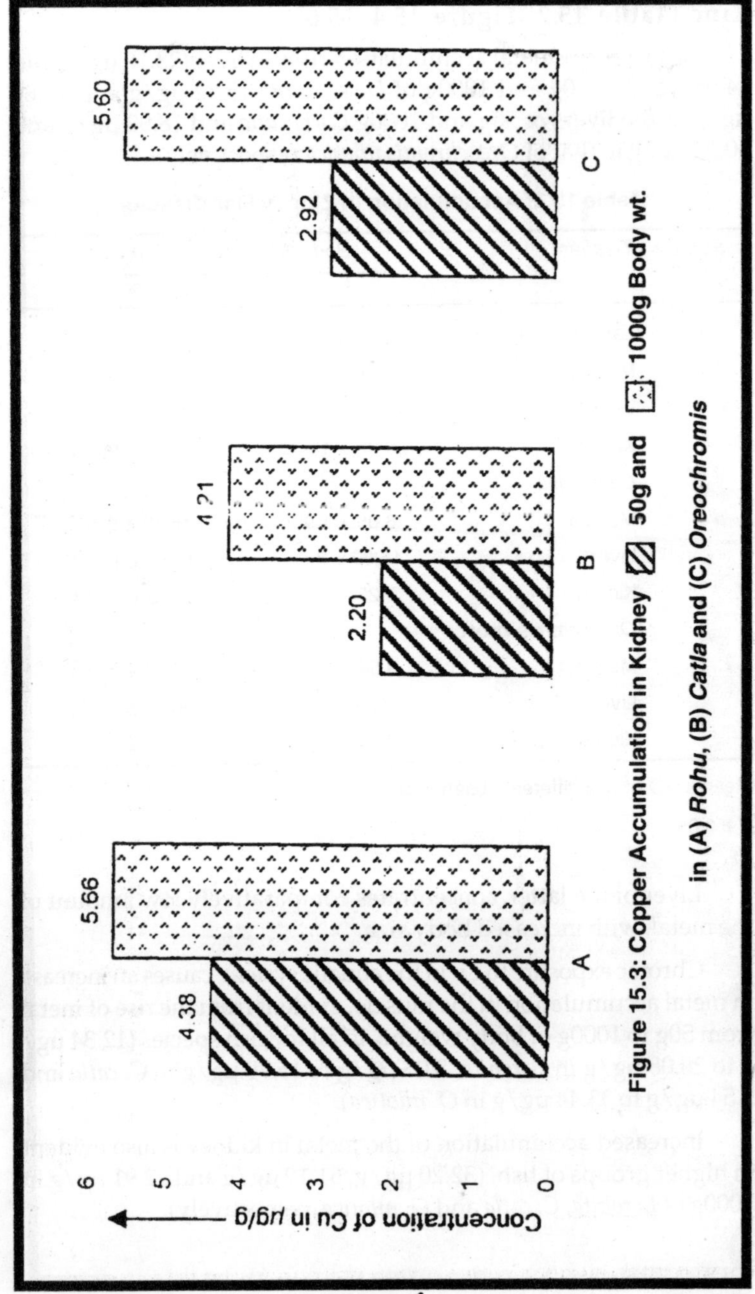

Figure 15.3: Copper Accumulation in Kidney ▨ 50g and ⛆ 1000g Body wt. in (A) *Rohu*, (B) *Catla* and (C) *Oreochromis*

Zinc (Table 15.2; Figure 15.4–15.6)

With age the metal accumulation rises in liver (28.10 µg/g and 44.96 µg/g in 50g and 1000g of *Labeo rohita* 15.82 µg/g and 46.80 µg/g in the livers of 50g and 1000g *Catla catla* and 18.92 µg/g and 30.57 µg/g in that of *Oreochromis nilotica* respectively).

Table 15.2: Accumulation of Zinc in Fish Tissues

Sl.No.	*Tissues*	*Body Weight of Fish (g)*	
		50	*1000*
1.	*Labeo rohita*		
	Muscle	12.34 ± 5.33	26.08 ± 6.70
	Liver	28.10 ± 11.02	44.96 ± 10.02
	Kidney	26.26 ± 10.02	32.20 ± 11.05
2.	*Catla catla*		
	Muscle	8.05 ± 1.90	15.79 ± 2.10
	Liver	15.82 ± 3.25	46.80 ± 13.20
	Kidney	7.25 ± 2.63	51.10 ± 14.13
3.	*Oreochromis nilotica*		
	Muscle	9.54 + 3.20	33.44 + 8.85
	Liver	18.92 ± 3.67	30.57 ± 6.49
	Kidney	10.78 ± 2.98	22.91 ± 5.46

Mean ± SD of five different observation.

$P < 0.05$.

Liver of the latter, concentrates comparatively low amount of the metal, with increased body size.

Chronic exposure to Zn in the aquatic system, causes an increase in metal accumulation in the muscle, evident from the rise of metal from 50g to 1000g of body weights in all the fish species (12.34 µg/g to 26.08 µg/g in *L. rohita*, 8.05 µg/g to 15.79 µg/g in *C. catla* and 9.54 µg/g to 33.44 µg/g in *O. nilotica*).

Increased accumulation of the metal in kidney is also evident in higher groups of fish. (32.20 µg/g, 51.10 µg/g and 22.91 µg/g in 1000g of *L. rohita*, *C. catla* and *O. nilotica*, respectively).

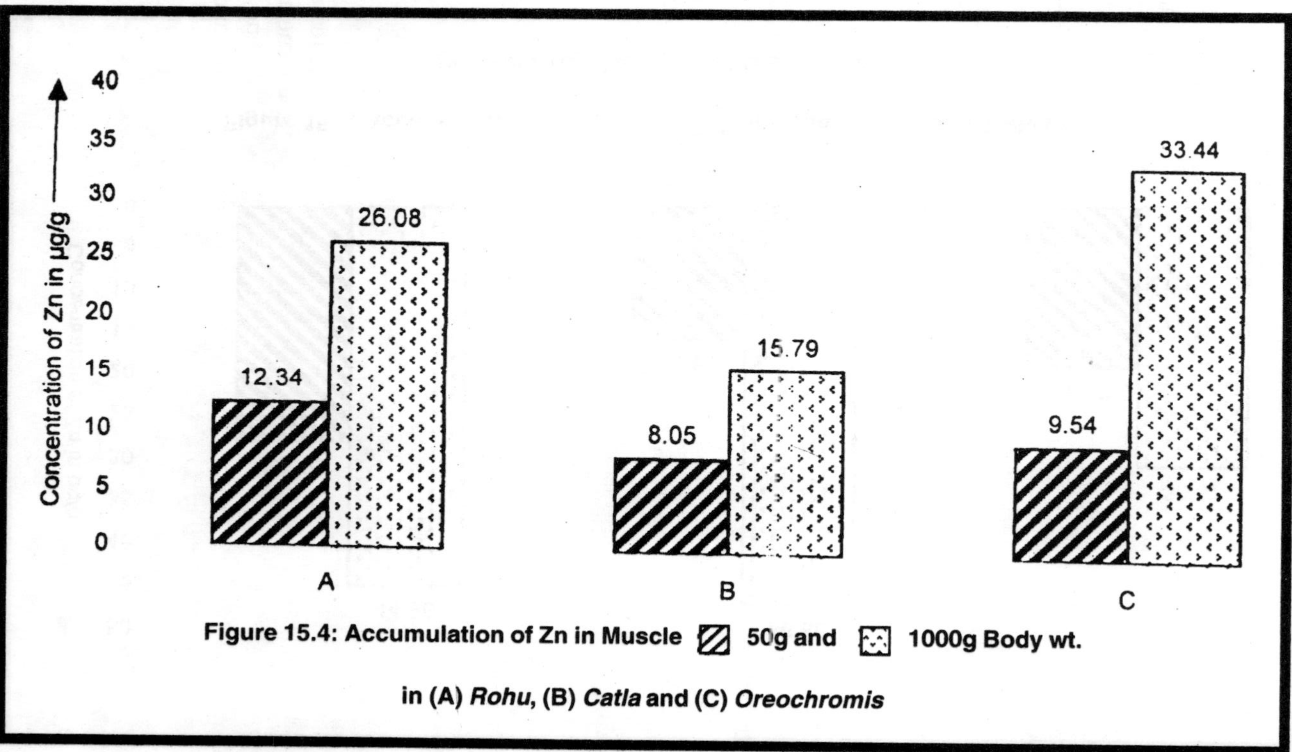

Figure 15.4: Accumulation of Zn in Muscle ▨ 50g and ⛶ 1000g Body wt. in (A) *Rohu*, (B) *Catla* and (C) *Oreochromis*

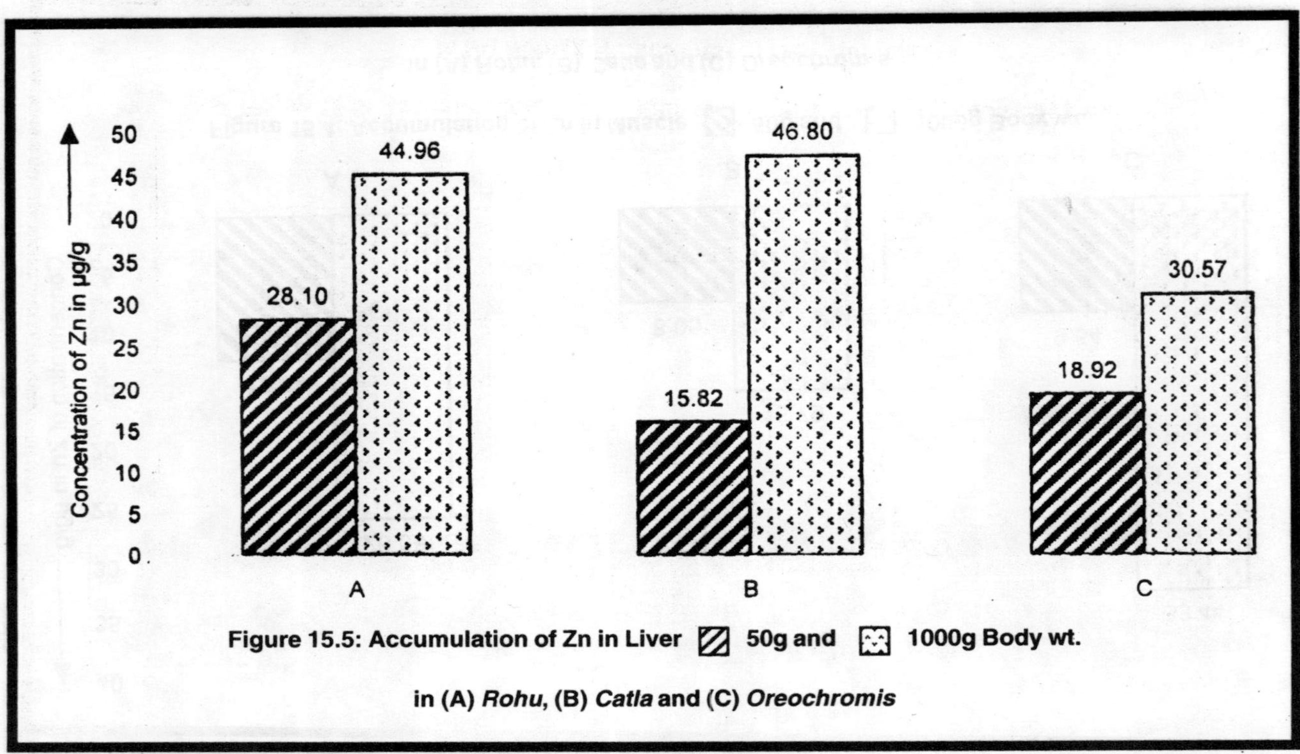

Figure 15.5: Accumulation of Zn in Liver ▨ 50g and ▩ 1000g Body wt. in (A) *Rohu*, (B) *Catla* and (C) *Oreochromis*

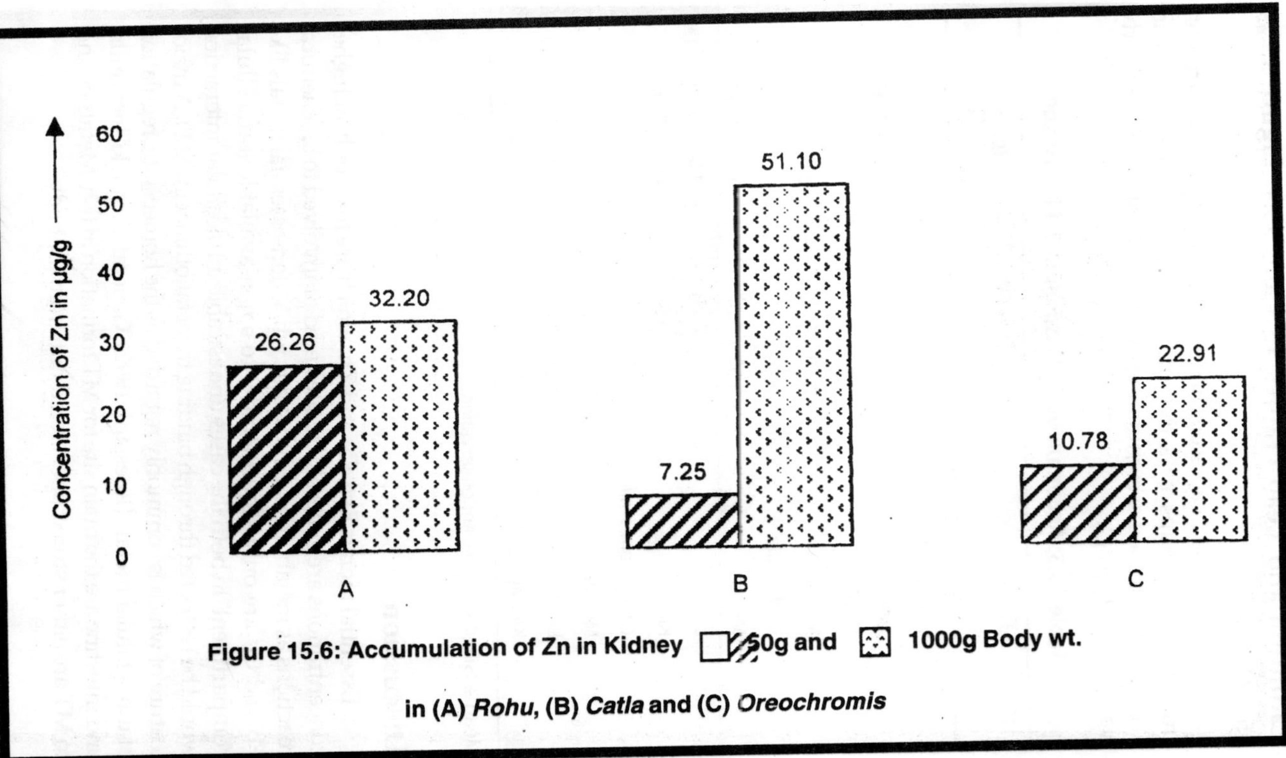

Figure 15.6: Accumulation of Zn in Kidney ▨ **50g and** ▩ **1000g Body wt. in (A) *Rohu*, (B) *Catla* and (C) *Oreochromis***

Cadmium and Lead (Tables 15.3 and 15.4, Figure 15.7–15.12)

Cadmium and Lead accumulation show insignificant variation among the age groups of the three different fish species. However, in *C. catla*, an increased accumulation of Pb is evident in 1000g of body weight (61.29 μg/g).

Table 15.3: Accumulation of Cadmium in Fish Tissue

Sl.No.	*Tissues*	*Body Weight of Fish (g)*	
		50	*1000*
1.	*Labeo rohita*		
	Muscle	3.31 ± 0.30	3.64 ± 1.00
	Liver	3.27 + 0.50	3.51 + 1.01
	Kidney	3.58 ± 0.72	3.50 ± 1.50
2.	*Catla catla*		
	Muscle	3.35 ± 0.52	3.43 ± 0.80
	Liver	3.52 ± 0.93	3.50 ± 0.99
	Kidney	3.18 ± 0.51	4.04 ± 0.56
3.	*Oreochromis nilotica*		
	Muscle	3.45 ± 0.35	3.07 ± 0.29
	Liver	3.51 ± 0.42	3.77 ± 0.50
	Kidney	4.01 ± 0.44	4.21 ± 0.22

Mean ± SD of five different observation.

Discussion

Essential metals like, Cu and Zn, which are toxic only at higher concentrations are generally sequestered or involved to biochemical reactions as metabolic needs dictate, while non-essential metals like Pb and Cd are only sequestered into a non-available intracellular compartment. In both the cases undesirable intracellular interaction would be restricted through binding to metallothionein (MT). A major feature of which is commonly regarded as the homeostatic regulation of intracellular metals (Roesjadi, 1992). Generally liver, kidney, gills, and intestine are the main site for MT formation in fish. Metals bound to MT are either stored or excreted out, from the body.

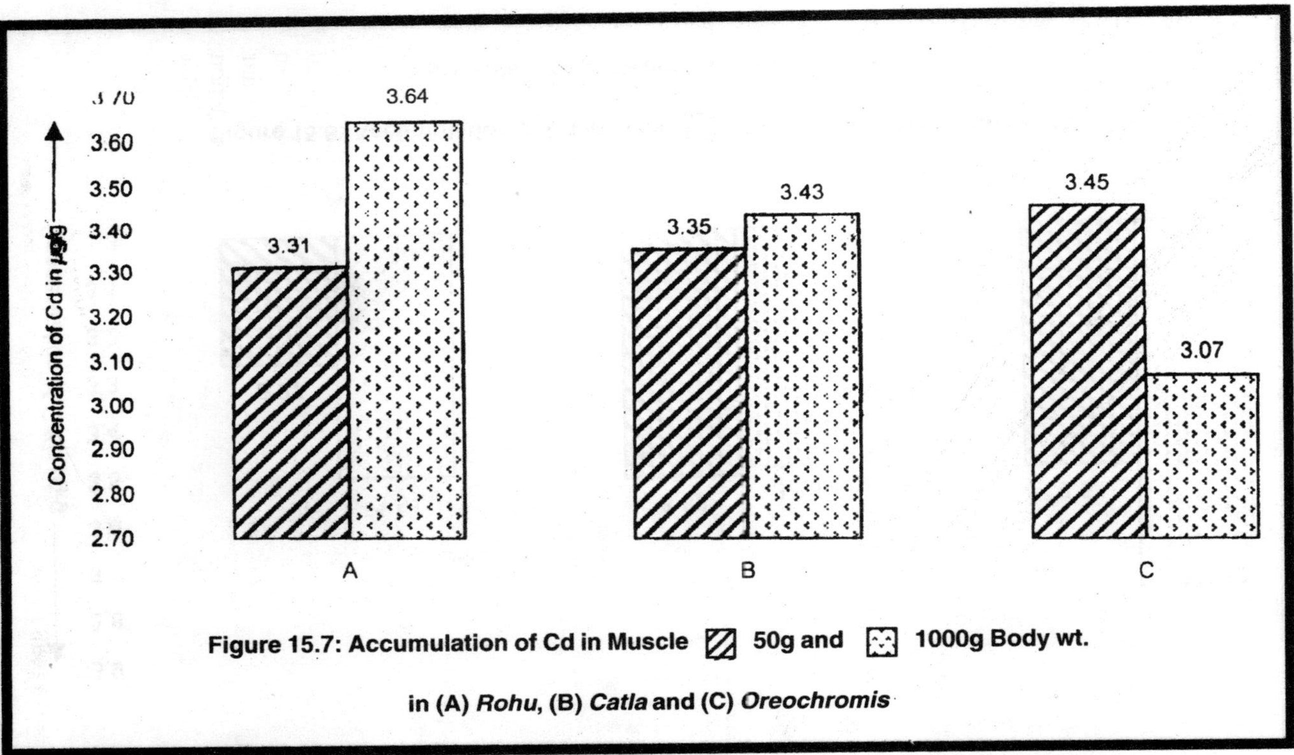

Figure 15.7: Accumulation of Cd in Muscle ▨ 50g and ⬚ 1000g Body wt. in (A) *Rohu*, (B) *Catla* and (C) *Oreochromis*

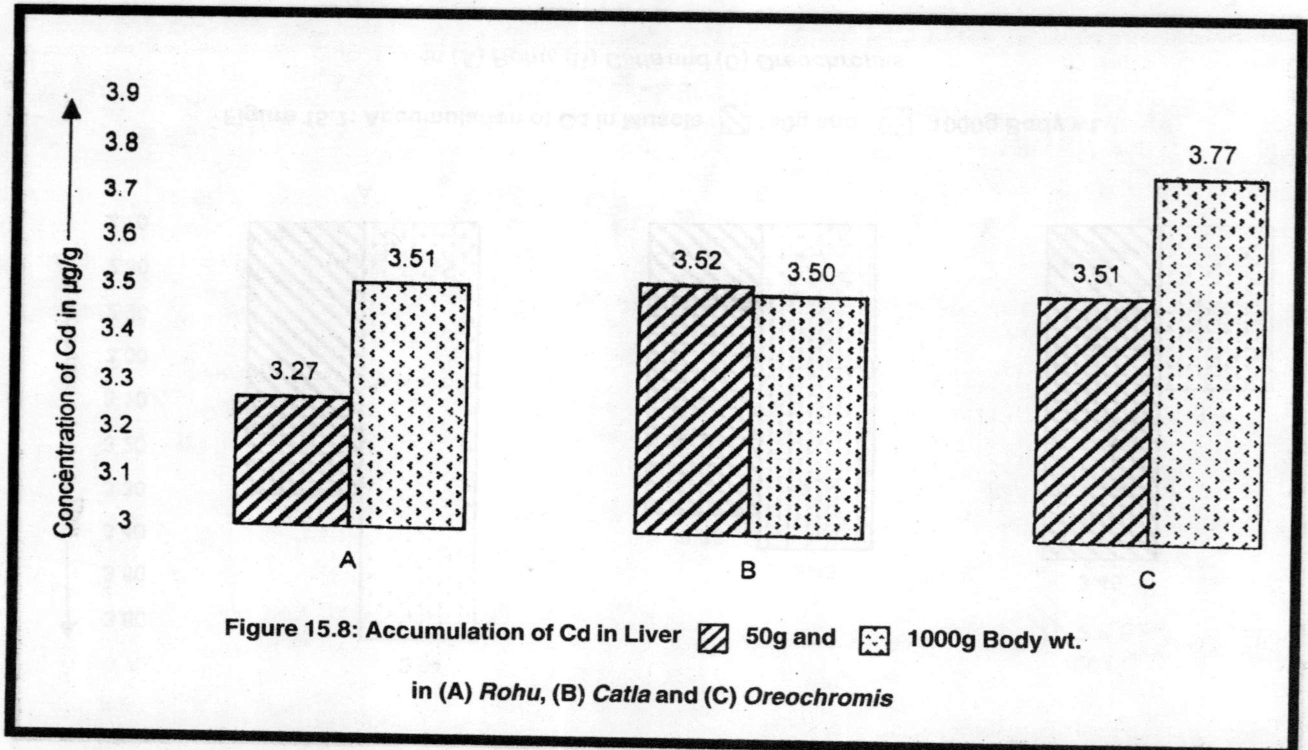

Figure 15.8: Accumulation of Cd in Liver ▨ 50g and ▩ 1000g Body wt. in (A) *Rohu*, (B) *Catla* and (C) *Oreochromis*

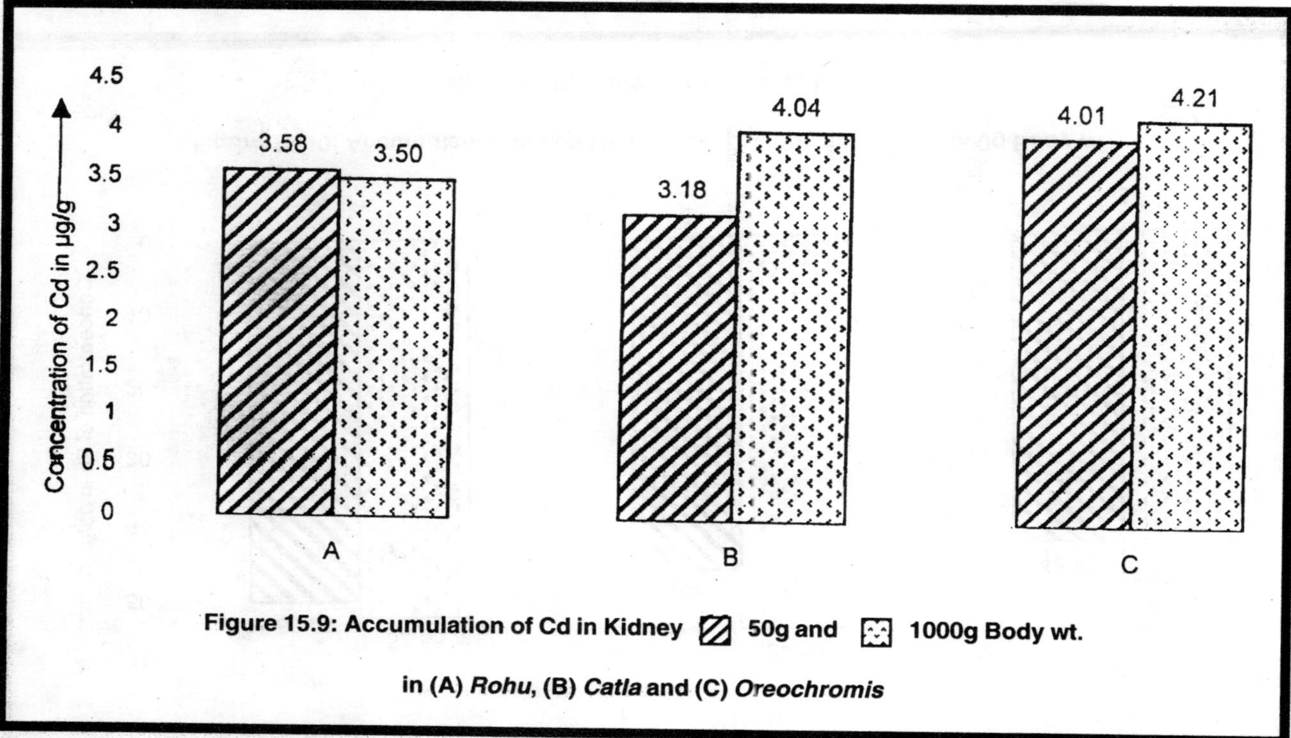

Figure 15.9: Accumulation of Cd in Kidney 50g and 1000g Body wt. in (A) *Rohu*, (B) *Catla* and (C) *Oreochromis*

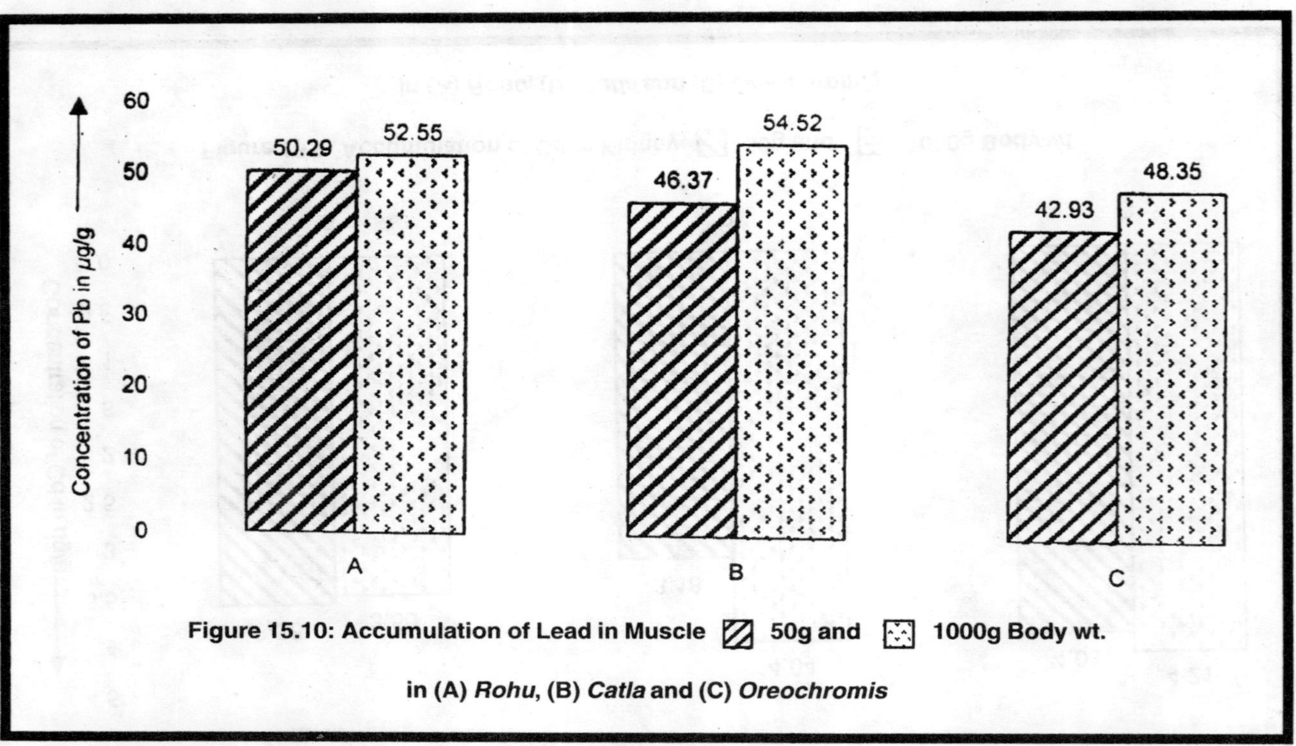

Figure 15.10: Accumulation of Lead in Muscle 50g and 1000g Body wt. in (A) *Rohu*, (B) *Catla* and (C) *Oreochromis*

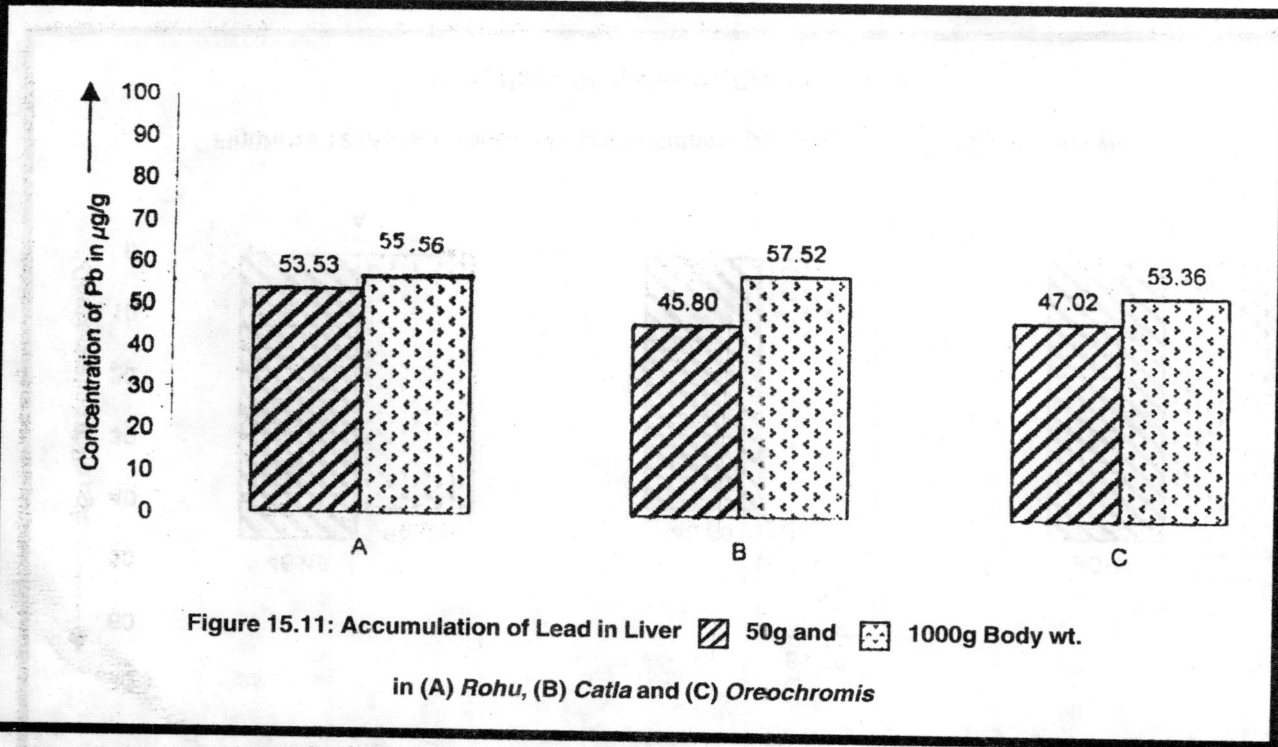

Figure 15.11: Accumulation of Lead in Liver ▨ 50g and ⛆ 1000g Body wt. in (A) *Rohu*, (B) *Catla* and (C) *Oreochromis*

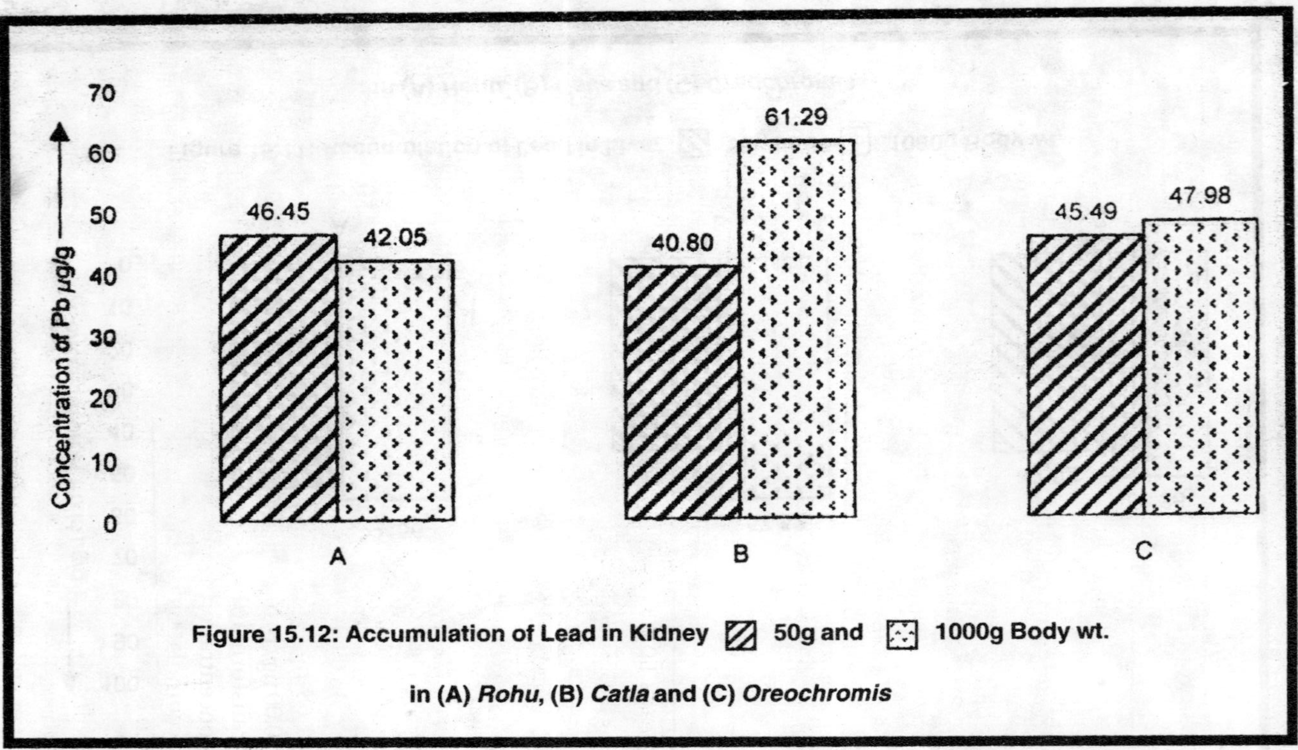

Figure 15.12: Accumulation of Lead in Kidney ▨ 50g and ⬚ 1000g Body wt. in (A) *Rohu*, (B) *Catla* and (C) *Oreochromis*

Table 15.4: Accumulation of Lead in Fish Tissues

Sl.No.	*Tissues*	*Body Weight of Fish (g)*	
		50	*1000*
1.	*Labeo rohita*		
	Muscle	50.29 ± 2.93	52.55 ± 9.00
	Liver	53.53 ± 5.30	55.56 ± 7.74
	Kidney	46.45 + 2.43	42.05 + 8.02
2.	*Catla catla*		
	Muscle	46.37 ± 4.76	54.52 ± 8.54
	Liver	45.80 ± 4.77	51.52 ± 8.54
	Kidney	40.80 ± 9.48	61.29 + 9.50
3.	*Oreochromis nilotica*		
	Muscle	42.93 ± 5.00	48.35 + 6.02
	Liver	47.02 ± 8.32	53.36 + 4.44
	Kidney	45.49 + 7.06	47.98 + 5.03

Mean ± SD of five different observation.

$P < 0.05$.

Metals are mostly absorbed in the body through gills, skin and food. Gills, skin and kidney act the major excretory route in fish. Dietary exposure of metals, to fish is however less toxic. Acute levels of metals in water stimulate mucus secretion which serve as a way of reducing their uptake, (Part and Lock, 1983). At chronic exposure, mucus secretion decreases, facilitating their uptake.

In the present study Cu, accumulation is more in liver (*L. rohita*–61.68 µg/g, *C. catla*–20.69 µg/g, and *O. nilotica*, accumulating 90.00 µg/g) because it acts as the main site for storage and metabolic transformation of metals. With Cu, the organ has the highest concentration factor for almost any exposure, where the actual accumulation depend on the nutritional status of the fish. Dietary Cu is directly taken in the liver from the intestine while water borne Cu is initially accumulated in the various tissues. With chronic exposure, it is rapidly removed by the liver (Buckley *et al.*, 1982) where it is finally accumulated. Tnis can be the reason for its rise in liver and fall in muscle, (800 g, body weight of *L. rohita, C. catla* and *O. nilotica* accumulate only 3.64 µg/g and 3.59 µg/g in muscle, respectively). Thus metal level in muscle can be homeostatically controlled. Low levels of Cu in kidney, suggest the low capacity of the organ for detoxification and elimination of the metal (800 g, body weight of *L. rohita, C. catla* and *O. nilotica* accumulate only 5.66 µg/g, 4.21 µg/g and 5.60 µg/g in kidney).

Zn uptake is more in low pH. In fish, although there is no such organ for storage of the metal, yet liver and kidney show an increased accumulation possibly due to the formation of MT in them. However, the metal is a weak inducer of the protein. Fish inhabiting Zn contaminated water, accumulate about 40 per cent of the metal in liver (Hogstrand, 1994). In the present study, 800 g, of *L. rohita, C. catla* and *O. nilotica,* liver accumulate 44.96 µg/g, 46.80 µg/g and 30.57 µg/g of Zn respectively. Liver and kidney also act as excretory organ for the metal (Wicklund, 1991). In kidney variation of Zn accumulation reflect the pattern of the metal excretion through this organ. Dietary Zn is however excreted mainly from gills. Thus in a highly polluted aquatic system, longer and elevated Zn exposure, results in the rise of Zn in muscle, liver and kidney in fish with higher body weights (Table 15.2), because with chronic and continued exposure of the metal in water, metal uptake in tissues increase over its elimination. According to Newmann and Mitz, (1988) fish with higher body weights, have lower capacity to depurate metals.

Pattern of Cd and Pb accumulation do not vary significantly in the different tissues of the fish under study (Maiti and Banerjee, 1999). Although according to some authors (Handy, 1992; Calamari, 1982) liver and kidney are the main site of storage for these metals. Low level of Cd induce MT formations but with chronic exposure, Cd binding capacity of MT decreases. Thus low accumulation of the metal in the tissues is possibly due to the binding of the metal with thionein molecules with its subsequent depuration form the organs. Cd uptake is non-linear and the accumulation factor decreases with its increasing concentration in water (Pelgrom *et al.*, 1994). Once accumulated, the metal is slowly eliminated from the body through gills, skin and kidney. Kidney however appears to have far less ability to detoxify the metal than either liver or intestine. According to Olsson, (1989), constant level of metals in the tissues in various age groups of fish reflect high toxicant level in the blood content of the organs. Turnover rate of Cd is very slow resulting in body burdens, that can be cumulative over an individual's life time and this can be the reason behind the low accumulation of the metals (Tables 15.3 and 15.4). Murphy *et al.* (1978) opines that whole body concentration of Cd were not correlated with body sizes in some fish.

Pb, do not induce MT formation in tissues. In the present study, Pb accumulation do not vary significantly with age which is in contradiction with the observation of Spry and Wiener, (1991) who opines that liver, kidney and scales are the main site for Pb accumulation. *Catla* exihibits strong tendency to retain lead in the kidney. The metals are excreted mainly through skin mucus (Varanasi and Markey, 1978).

Accumulation of metals in the different components of the food chain, depend on the availability of the metal in both water and sediment. Difference in bioaccumulation in the tissues of the organism is dependent on their physiological activities.

In the present study, *Orcochromis nilotica* (90.00 µg/g) show an increased accumulation of Cu in liver, than either *Labeo rohita* (61.68 µg/g) or *Catla catla* (20.69 µg/g). It seemed that nutritional level and position of the fish in the trophic strata, determine the accumulation pattern and toxicity of the metals. (Tao, S. *et al.*, 1999; Kraal, *et al.*, 1995) *Oreochromis* sp feeding on a wider variety food both from planktons to the detritus from the pond bottom, concentrates more Cu in its liver. Studies reveal that an excess of Cu is always found in the sediments, than in water. The detoxification of Cu, may be by sequestering and storage in *O. nilotica,* rather than elimination from liver (Giesy and Wiener, 1977).

Zinc too shows higher accumulation with age, in muscle, liver and kidney of all the three fish species under study. However hepatic tissues (46.80 µg/g) and kidney (51.10 µg/g) of *Catla catla,* accumulate the most. Generally dietary Zn is not excreted through kidney, so it may be assumed that the metal is only accumulated in the organ, which plays little role in its excretion.

Insignificant variation in accumulation of essential metals like Cd and Pb (Tables 15.4 and 15.5) in the different tissues, of the three fish species suggest that feeding habit and trophic status exert little influence on the accumulation of these metals.

Toxicity of metals, does not solely depend on its accumulation in the tissues. Mechitsuka and Coworkers, (1987) suggested that toxicity of Cd depend on the accumulation of Cu, as it displaces Cd from the metallothionein molecules. Scheuhammer and Cherian, 1986 opines that during Cu exposure, Cd binding capacity of metal decreases with subsequent elimination of Cd. Cu exposure also affect faecal excretion of Cd (Pelgrom *et al.*, 1994). On the other hand, synergistic effect of Pb on Cu uptake and elimination and vice versa was observed by Tao *et al.* (1999). Zn intake is however inhibited by low levels of Cd, but Zn seems to have no effect on Cd uptake suggesting complex interaction of these metals in nature.

Acknowledgement

The first author is grateful to CSIR for financial support.

References

Buckley, J.T., Roch, M., McCarter, J.A., Rendel, C.A. and A.T. Mathiieson (1982). Chronic exposure of coho salmon to sublethal concentration of Copper. I. Effect on growth, on accumulation and distribution of copper and on copper tolerance. *Comp. Biochem. Physiol.*, 72C(1): 15–19.

Calamari, D., Gaggino, F.F. and G. Pacchetti (1982). Toxico-kinetics of low levels of Cd, G, Ni and their mixture in long term treatment on *Salmo gaidneri* Rich. *Chemosphere,* 11: 59–70.

Chernoff, B. (1975). A method for wet digestion of fish tissue for heavy metal analysis. *Trans. Am. Fish.*, 104: 803–804.

Giesy, J.P. Jr. and J.G. Wiener (1977). Frequency distribution of trace metal concentrations in five fresh water fishes. *Trans. Am. Fish. Soc.*, 106: 893–403.

Gjessing, E.T. (1981). The effect of aquatic humus on the biological availability of cadmium. *Arch. Hydrobiol.*, 91: 144–149.

Grahl, K., Franke, P. and R. Hallebach (1985). The excretion of heavy metals by fish. *Symp. Biol. Hung.*, 29: 357–364.

Graney, R.L., Cherry, Jr. D.S. and Jr. J. Cairns (1984). The influence of substrate pH, diet and temperature upon cadmium accumulation in the asiatic clam (*Corbicula fluminae*) in laboratory and in artificial streams. *Water Res.*, 18: 833–842.

Gupta, S.K. and A. Mitra (1997). Utilization of solid and liquid wastes of Kolkata city for Agricultural uses. *J. Ind. Soc. Soil Sc.*, p. 52–56.

Hamilton, S.J. and Mehrle, P.M. (1986) Metallothionein in Fish: Review of its importance in assessing stress from metal contaminants. *Trans. Am. Fish. Soc.*, 115: 596–609.

Handy, R.D. (1992). The assessment of episodic metal pollution. II. The effects of cadmium and copper enriched diets on tissue contaminant analysis in Rainbow Trout (*Oncorhynchus mykiss*). *Arch. Environ. Contam. Toxicol.*, 22: 82–87.

Heath, A.G. (1995). *Water Pollution and Fish Physiological.* CRC Press, NC. Florida, USA.

Hogstrand, C., Wilson, R.W., Polgar, D. and C.M. Wood (1994). Effects of Zn on the branchial calcium uptake in freshwater rainbow trout during adaptation to waterborne zinc. *J. Exp. Biol.*, 186: 55–73.

Kotze, P., Du Preez, H. H. and J.H.J. Van Vuren (1999). Bioaccumulation of copper and zinc in *Oreochromis mossambicus* and *Clarias gariepinus* from the Olifants River, Mpumalanga, South Africa. *Water S.A.*, 25(1): 99–110.

Kraal, M.H., Kraal, M.H.S., De-Groot, C.J. and C. Davids (1995). Uptake and tissue distribution of dietary and aqueous cadmium by carp (*Cyprinus carpio*). *Ecotoxicol. Environ. Saf.*, 31(2): 179–183.

Lliang, Y., Cheung, R.Y.H. and M.H. Wong (1999). Reclamation of wastewater for polyculture of freshwater fish: Bioaccumulation of trace metal in fish. *Wat. Res.*, 33(11): 2690–2700.

Maiti, P. and S. Banerjee (1999). Accumulation of heavy metals in different tissues of the fish, *Oreochromis nilotica*, exposed to wastewater. *Env. & Ecol.*, 17(4): 895–898.

Meshitsuka, S., Ishizawa, M. and T. Nose (1987). Uptake and toxic effects of heavy metal ions: interactions among cadmium, copper and size in cultured cells. *Experientia*, 43: 151–156.

Moriarty F., Hanson, H.M. and P. Freestone (1984). Limitation of body burden as an index of environmental contamination heavy metal in fish *Cottus gobio* L. from the River Ecclesbourne, Derbyshire. *Environ. Pollut. Ser.*, A34: 297–320.

Murphy, B.R., Atchison, G.J., McIntosh, A.W. and D.J. Kolar (1978). Cadmuim and Zinc Content of fish from an industrially contaminated lake. *J. Fish. Biol.* 13: 327–335.

Newman, M.C. and McIntosh, A.W. (1983). Lead elimination and size effects on accumulation by two freshwater gastropods. *Arch. Environ. Contam. Toxical.*, 12: 25–59.

Newmann, M.C. and S.V. Mitz (1988). Size dependence of Zn elimination and uptake from water by mosquito fish *Gambusia affinis. Aquat. Toxicol.*, 12: 17–32.

Olsson, P.E., Zafarullah, M., Foster, R., Hamor, T. and L. Gedamu (1990). Developmental regulation of metallothionein mRNA, zinc and copper levels in rainbow trout, *Salmo gairdneri. European J. Biochem.*, 193: 229–236.

Part, P. and R.A.C. Lock (1983). Diffusion of Ca, Cd, and Hg in a mucous solution from rainbow trout. *Comp. Biochem. Physiol.*, 76C: 259–263.

Pelgrom, S.M.G.J., Lamers, L.P.M., Garritsen, J.A.M., Pels, B.M., Lock, R.A.C., Balm, P.H.M. and S.E. Wendelaar Bonga (1994). Interactions between copper and cadmium during single and combined exposure in juvenile tilapia, *Oreochromis mossambicus.* Influence of feeding condition on whole body metal accumulation and the effect of the metals on tissue water and ion content. *Aquatic Toxicology*, 30: 117–135.

Roesijadi, G. and P. Klerk (1989). A Kinetic analysis of Cd finding metallothionein and intra cellular ligands in oyester gills. *J. Exp. Zool.*, 251: 1–12.

Roesjadi, G. (1992). Metallothioneins in metal regulation and toxicity in aquatic, animals. *Aquatic Toxicology*, 22: 81–114.

Scheuhammer, A.M. and M.G. Cherian (1986). Quantification of metallothioneins by a silver saturation method. *Toxicol. Appl. Pharmacol.*, 82: 417–425.

Spry, D. and J. Wiener (1991). Metal bioavailability and toxicity to fish in low alkalinity lakes: A critical review, *Environ. Pollut.*, 71: 2443–253.

Tao, S. Liang, T., Cao, J., Dawson, R.W. and C. Liu (1999). Synergistic effect of Cu and Pb uptake by Fish. *Ecotoxical. & Environ Saf.*, 44: 190–195.

Unlu, M.Y. and S.W. Fowler (1997). Factors affecting the flux of arsenic through the mussel, *Mytilus gallovincialis, Mar. Biol.*, 51: 209–219.

Varanasi, U. and D. Markey (1978). Uptake and release of lead and cadmium in skin and mucus of coho salmon (*Oncorhynchus kisutch*). *Comp. Biochem. Physiol.*, 60C: 187–191.

Wicklund Glynn, A. (1991) Cadmium and Zinc Kinetics in fish: studies on water borne ^{100}Cd and ^{65}Zn turnover and intracellular distribution in minnows, Phoxinus phoxinus. *Pharmacology & Toxicology*, 69: 485–491.

Chapter 16

Cadmium Chloride Effects on Thyroid and Gonads of the Fish *Channa orientalis* (Sch.)

S.V. Deshmukh and K.M. Kulkarni

ABSTRACT

Thyroid gland of cadmium chloride treated Channa orientalis showed marked stimulatory effect i.e., hypertrophy but less secretion and hyperplasia in follicular cells.

Cadmium chloride exposed C. orientalis showed retardation of gonadal maturation and brought about retardation in testicular activities and degeneration of testes.

The testicular size decreases, spermatogenic cells became atrophic i.e., spermatogonia and spermatocytes lost their characteristics and reduced in number while interstitial cells showed hypertrophied nature.

Keywords: *Thyroid gland, Cadmium chloride, Antispermatogenic, Channa orientalis, Testes, Degeneration.*

Introduction

Since the last decade several investigations were undertaken to study the role of non-steroidal compounds in the control of reproduction in rodents and other animals. The use of these substances have given striking results especially in controlling or interfering the process of spermatogenesis. Cadmium chloride is one such compound which is actually given in an appropriate physiological dosages proved to be antispermatogenic, which not only suppresses spermatogenesis but also destroys the leydig cells of the testes. Perusals of the literature pertaining to the use of this compound suggests that the drug has been studied especially in relation to its role in bringing about changes in pituitary gland and gonads. However its action on thyroid and other gland has been ignored. The mechanism by which cadmium chloride exerts action on different endocrine gland is not fully assessed and need further investigations. It is therefore decided to investigate the effects of cadmium chloride on the thyroid and gonads of freshwater fish *Channa orientalis.*

Thyroid Gland

The thyroid gland is affected by mercuric chloride, ammonia, phenol etc. and influences the thyroid function (Bhattacharya *et al.*, 1989). However to study the exact effect of cadmium chloride on thyroid gland of fish *C. orientalis*, the animals treated with the same.

Gonads

The gonads were affected by cadmium chloride which is chemosterilent and also known for its toxic properties. After administration of cadmium chloride the animals showed a decline in the weight of their body as well as of the testes (Sharma *et al.*, 1980). Some workers Van campoenhout, 1956; Parizek, 1957 and 1960; Kar *et al.*, 1958; Gunn *et al.*, 1961; Allanson and Deansely, 1962; Gunn and Gould, 1970 and Johnson, 1977, have demonstrated the degenerative changes in rat testes after cadmium chloride treatment. According to these workers testes injury after cadmium chloride treatment was followed by testicular necrosis, hemorrhage, cessation of spermatogenesis. Similar changes were also reported in other mammalian species. (Mouse-Parizek, 1957; Meck, 1959; Hamstar and Guinea pig-Parizek, 1960; Girod *et al.*, 1976; Rabbit-Cameron and Foster 1963; Gerbil-Ramaswamy and Kaul, 1966; Barliner and Jones-witters, 1975; Non-scrotal bat, R. Kinneri-Dixit and Lohiya, 1974; Singwi and Lal, 1980).

Perusal of literature indicates that the mechanism of action of this compound male gonads is not fully understand, so in the present investigation an attempt has been made to study the exact effect of cadmium chloride on male gonads (testes) of the fish *Channa orientalis*. For this purpose the animals were treated with the same.

Materials and Methods

Young and sexually mature *Channa orientalis* were selected and acclimatized to the laboratory condition.

For present investigation two concentration of cadmium chloride were tried. Two hundred and twenty four young, mature and acclimatized fished were selected having similar sizes and weights. The animals were divided into three groups.

Group I

(Total 112 fishes) as control group. Which is further subdivided into group I A and Group I B having 56 fishes in each group. Fishes were reared in test water.

Group II

As high concentration group for cadmium chloride in which fishes were treated in high concentration (15 mg/l) of cadmium chloride water (Total 56 fishes).

Group III

As low concentration group for cadmium chloride in which fishes were treated in low concentration 5 mg/l of cadmium chloride water (Total 5 fishes).

Group I A and Group I B were respectively used as control for group II and group III. For studying histological changes the fishes from group I, II and III were sacrificed by decapitation at the interval of 24, 48, 72, 96, 168, 360 and 720 hours of treatment. At sacrifice the different tissue like gonads, and lower jaws with thyroid were taken out carefully, washed with distilled water and fixed in Boin's fluid for 24 hours. The lower jaws were decalcified in 6 per cent solution of formic acid in 5 per cent formaldehyde for about a week. The tissues were dehydrated by passing through various grades of alcohol, cleared in xylene and a embedded in paraffin wax (56–58 c). Blocks were prepared and sectioned at 4–6 μ and stained with following techniques.

Gonads and lower jaw with thyroid were stained with Ehrilich haematoxylin–eosin technique.

Assessment of Thyroid Activity

Assessment of thyroid activity was measured by microscopic examination of transverse sections (5-6 μ) of lower jaws in both control and experimental groups. This was supplemented by measurements in the following parameters.

1. Average number of follicles in a, section (5 to 6 section/fish).
2. Average follicular cell height was calculated by measuring 5 to 6 cells par follicles at random in each of the 4 to 5 follicles per section. Thus measurements were carried out in 5 sections per animal. Thus the average epithelical cells height were calculated of animal's thyroid gland.
3. Average size of the follicles were estimated by measuring diameter of 5 to 6 follicles per section and 5 sections per animal. And thus the mean diameter of follicles was found out of the animal.
4. Average number of cells per follicles were counted by random follicles per section. This procedure was followed for 5 follicles per section and 5–6 slides per animal tissue, yielded the average number of cells per follicle.

Assessment of Gonads (Testes)

Assessment of the gonadal activity at microscopic level was based on the following parameters:

1. Gonadal diameter.
2. Differential cell count of germinal cells.

Observations on the histology of the testes of each animal relating both the groups including control and experimentals were based on the examination of serial transverse sections. For each testes, 5 to 6 sections per animal were selected. Their diameter were measured with the help of 1 mm linear ocular micrometer at magnification of 100 x–counting of the various testicular cell types observed in the sections were made at magnification of 450 x with the help of an ocular micrometer having 1 sq./cm. area divided into quadrants of 0.25m each. In each animal 5 sections were selected with five areas per sections. Thus the figure obtained were the average values per animal and in turn the mean values for a group of animal.

The data obtained after various measurements were tested for statistical significance. The experimental values were statistically evaluated by student's 't' formula.

Results and Observations

Thyroid Gland

Cadmium Chloride Treated (Experimental) Fish Thyroid

Histological Changes in High Concentration Exposure (15 mg/l) (Plate 16.1–A, B, C, D, E, F, Table 16.1, Figure 16.1)

Statistical analytic values of histological changes of thyroid gland are given in Table 16.1. After 24–48 hrs. exposure few follicles peripheral region showed degeneration. Most of the follicle became irregular in shape. The number of follicles per section was reduced from control value 48.096 to 39.116 upto 360–720 hours of exposure. The diameter of follicles were also found reduced. Its percent changed values at 24–48 h, 72–96 h, 168–360 h, 360–720 h were as 7.555, –7.994, –8.775, –9.994 respectively (Table 16.1) secretary cells showed maximum hypertrophy and hyperplasia. Numerical increase in cells per follicles were from 35.026 (control value) to 45.752/24–48 h, 47.170/72–96 h; 50.214/168–360 h; 55.084/360–720 h. Some degenerated follicles were also observed (Plate 16.1 D, E). In degenerated follicles, epithelial cells detached from follicular well and have accumulated in the lumen. Pycnotic nuclei were observed. Some follicles of peripheral region showed vacuolation in the colloidal material. Most of the secretory cells showed degranulation in their cytoplasm. Follicular wall of most of the follicles showed hypertrophy due to which it appeared to be thick. Height of the follicular cells increased from 4.031 μ (control value) to 7.152 μ hrs. upto 360–720 hours of exposure. The follicular cells appeared to be columnar with basal nuclei. After 168 hours of treatment follicular wall was found to be collapsed and less contains of secretion. No significant changes were observed in the vasculature of the gland.

Histological Changes in Low Concentration (5 mg/l) Exposure–(Plate 16.2–A, B, C, D, E, F and Table 16.2, Figure 16.2)

The number of follicles reduced from control value of 48. 085 to 43.369 upto 360–720 hours of exposure. Some irregular shaped and few empty follicles were observed. Percent changed for follicular diameter are recorded in the Table 16.2 as –4.088 (24–48 h): 4–0.332

Table 16.1: Effect of Cadmium Chloride Treatment on the Thyroid Gland (Histological Parameters) of the Fish *C. orientalis* (high concentration–15mg/l)

Experimental Period (h)	*Follicles Per Section*	*Cell Height (μ)*	*Follicular Diameter*	*Cells Per Follicles*
Control	48.096 ± 2.877	4.031 ± 0.143	60.354 ± 3.843	35.026 ± 3.296
24–48	41.819* ± 1.737	6.872* ± 0.187	56.256* ± 1.978	45.752* ± 1.881
	(–13.050)	(70.478)	(–7.555)	(30.622)
72–96	41.161* ± 1.603	7.003* ± 0.228	55.989* ± 1.865	47.170** ± 1.868
	(–14.419)	(73.728)	(–7.994)	(–34.671)
168–360	40.348* ± 1.414	7.054* ± 0.221	55.514** ± 1.821	50.214*** ± 1.837
	(–16.109)	(74.993)	(–8.775)	(–43.362)
360–720	39.116** ± 1.157	7.152* ± 0.230	54.772* ± 1.737	55.084 ± 1.774
	(–18.670)	(77.424)	(–9.994)	(–46.301)

P values: *: < 0.1; **: < 0.01; ***: <0.001.

Non Significant, ( ): Parenthesis figures are percentage change.

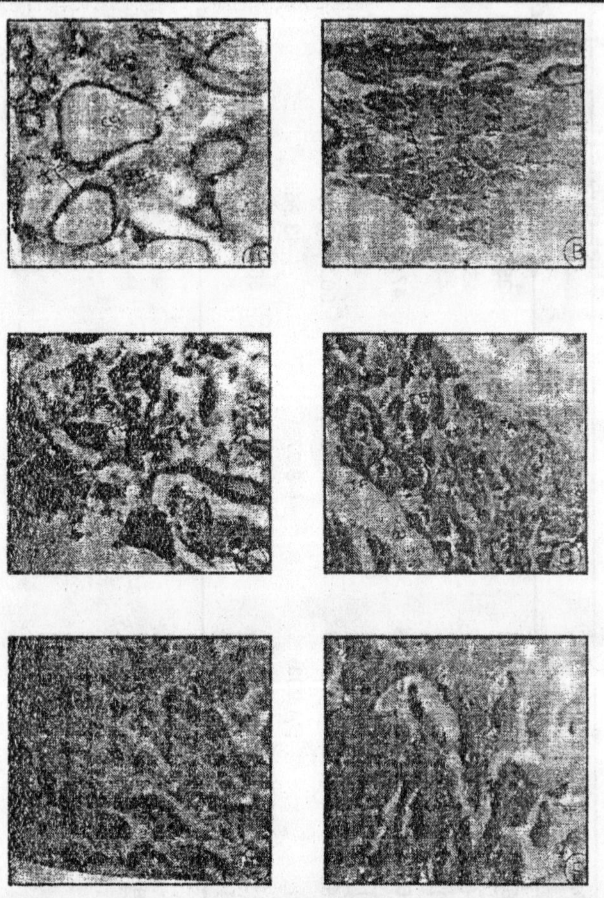

Plate 16.1: Histological Changes in Thyroid of *Channa orientalis* Exposed High Concentration of Cadmium Chloride

(A) T. S. of control *C. orientalis* thyroid; (B) T.S. of cadmium chloride exposed *C. orientalis* thyroid during early hours (24–48h.); (C) T.S. of cadmium chloride exposed *C. orientalis* thyroid during middle hours (72–96h); (D) T.S. of cadmium chloride exposed *C. orientalis* thyroid during middle hours (168h); (E) T.S. of cadmium chloride exposed *C. orientalis* thyroid during late hours (360h); (F) T.S. of cadmium chloride exposed *C. orientalis* thyroid during late hours (720h).

CO: Colloid; VC: Vacuolated Colloid. FE: Follicular epithelium; TF: Thyroid follicles; NU: Nucleus.

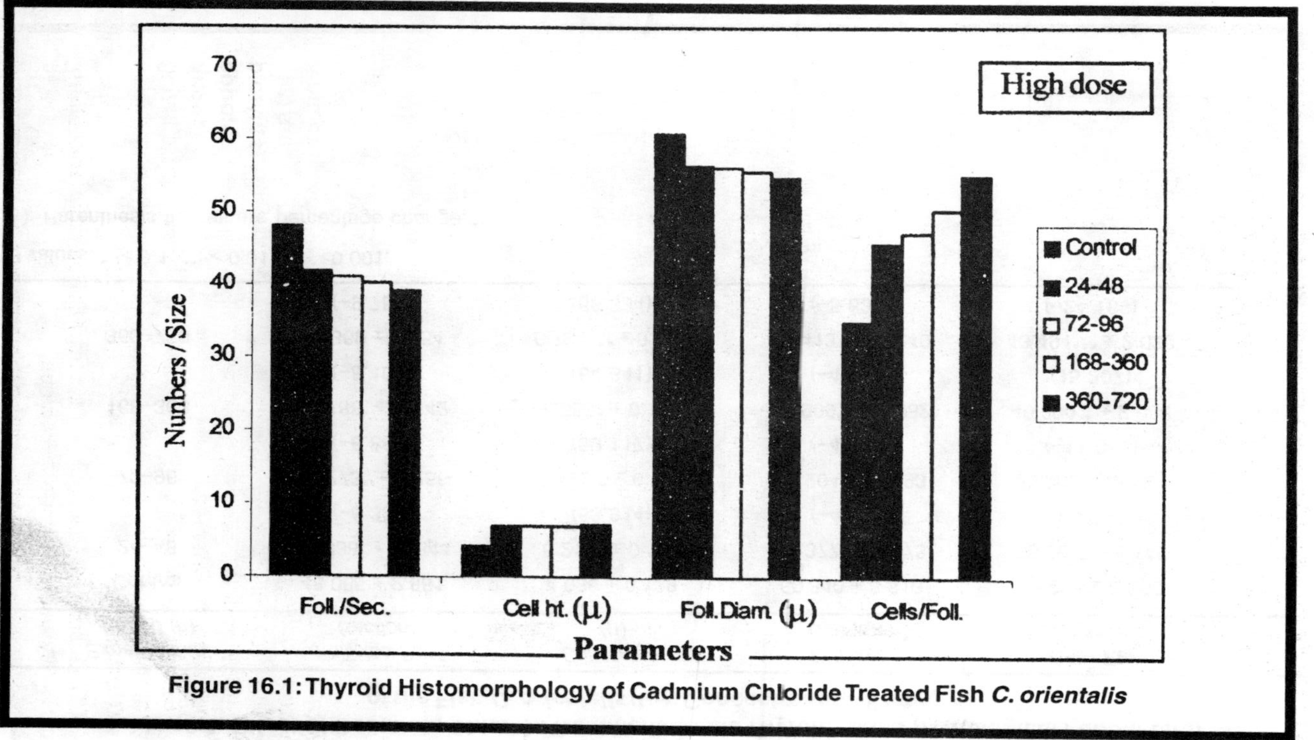

Figure 16.1: Thyroid Histomorphology of Cadmium Chloride Treated Fish *C. orientalis*

Table 16.2: Effect of Cadmium Chloride Treatment on the Thyroid Gland (Histological Parameters) of the Fish *C. orientalis* (low Concentration –5 mg/l)

Experimental Period (h)	*Follicles Per Section*	*Cell Height (μ)*	*Follicular Diameter*	*Cells Per Follicles*
Control	48.085 ± 2.861	4.036 ± 0.138	60.840 ± 3.816	35.028 ± 3.280
24–48	45.038* ± 2. 344	6.212* ± 0.164	58.377* ± 2.373	38.263* ± 2.145
	(–6.336)	(53.914)	(–4.048)	(–9.235)
72–96	44.773* ± 2.256	6.422** ± 0.175	58.204* ± 2.193	38.903* ± 2.121
	(–6.887)	(59.117)	(–4.332)	(–11.062)
168–360	44.188* ± 2.142	6.653* ± 0.186	57.909** ± 2.082	40.390** ± 2.083
	(–8.104)	(64.841)	(–4.817)	(15.307)
360–720	43.369 ± 1.954	6.794*** ± 0.204	57.413** ± 1.940	43.494*** ± 2.028
	(–9.751)	(68.334)	(–5.632)	(–24.169)

P values: *: < 0.1; **: < 0.01; ***: <0.001.

( ): Parenthesis figures are percentage change.

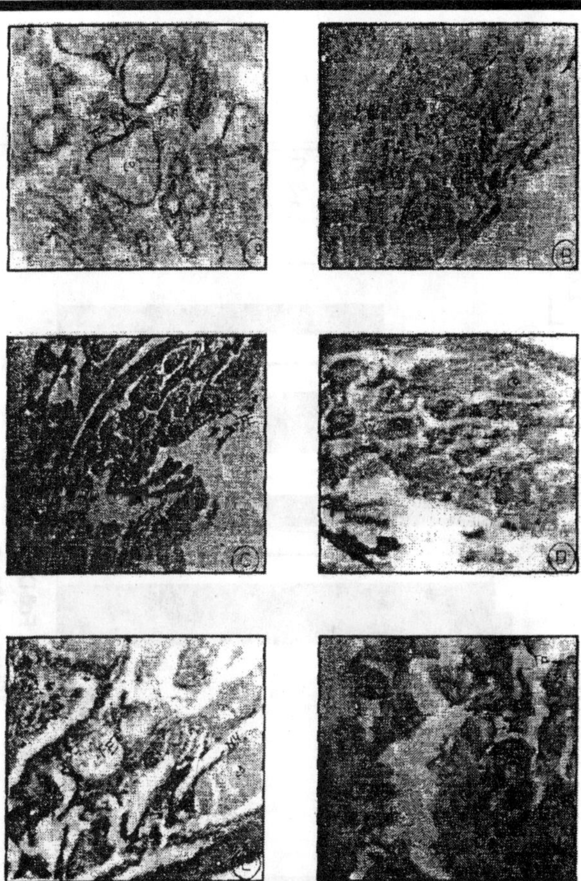

Plate 16.2: Histological Changes in Thyroid of *Channa orientalis* Exposed Low Concentration of Cadmium Chloride

(A) T. S. of control *C. orientalis* thyroid; (B) T.S. of cadmium chloride exposed *C. orientalis* thyroid during early hours (24–48h.); (C) T.S. of cadmium chloride exposed *C. orientalis* thyroid during middle hours (72–96h); (D) T.S. of cadmium chloride exposed *C. orientalis* thyroid during middle hours (168h); (E) T.S. of cadmium chloride exposed *C. orientalis* thyroid during late hours (360h); (F) T.S. of cadmium chloride exposed *C. orientalis* thyroid during late hours (720h).

CO: Colloid; VC: Vacuolated Colloid. FE: Follicular epithelium; TF: Thyroid follicles; NU: Nucleus.

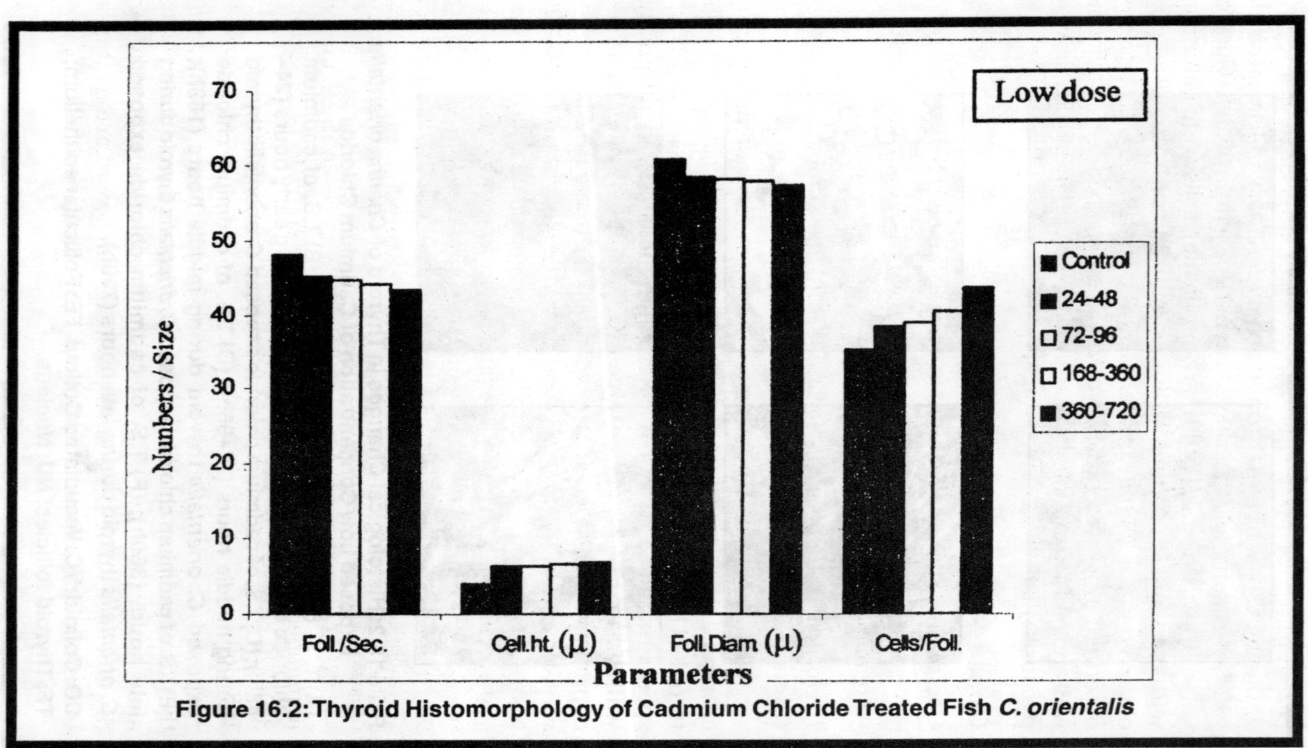

Figure 16.2: Thyroid Histomorphology of Cadmium Chloride Treated Fish *C. orientalis*

(72–96h); –4.817 (168–360 h); –5.632 (360–720h). Follicular epithelium showed hypertrophy and hyperplasia. Cells per follicle were also found to be increased numerically from 35.028 (Control value) to 43.494, (360–720 h). The follicular cells were columnar with basal nuclei, obliterated lumen and foldings. Vacuoles were observed in the colloidal material. Few degenerated follicles were observed, in which epithelial cells were detached from follicular wall and had accumulated in the lumen. Nuclei of these cells showed pycnosis (Plate 16.2–E, F).

Gonads (Testes)

Cadmium Chloride Treated (Experimental) Fish Testes

Statistical analysis of the histological structure, values of testes were recorded in Tables 16.3 and 16.4.

Histological Changes at High Concentration (15 mg/l) Exposures (Plate 16.3–A, B, C, D, E, F; Table 16.3, Figure 16.3)

Cadmium chloride treated fish testes showed reductional changes. Testicular diameter reduced from control value of 798.186µ to 651.211 µ upto 360–720 hours of treatment. Spermatocytes, spermatids and spermatozoa were found to be detached from the wall and accumulated in the lumen. Respective reduction in number of primary spermatogonia and secondary spermatogonia were from control value of 6.304 to 5.083 and from 7.843 to 5.707 upto 360–720 hours of treatment. Most of the spermatogonia and spermatocytes showed necrosis. Primary spermatocytes gradually decreased in number from 48.521 (Control value) to 43.611 (24–48h): 43.116 (72–96 H); 42.427 (168–360 h); 41.364, (360–720 h). Secondary spermatocytes were also found to be reduced in number. The observed values in percent change were –26.232, –26.502, –33.223, –42.043 at exposure period of 24–48 h; 72–96 h; 168–360 h. Reduction in number of spermatid was noticed from 191.392 to 140.132 upto 360–720 hours of exposure. Percent changes at 24–48 h; 72–96 h; 168–360 h; 360–720 h for number of spermatozoa were –6.414, –7.015, –8.353, –10.215 (Table 16.3). The degenerative changes in the germinal cells were marked by distruction of cell wall, nucleus and vacuolization in cytoplasm. Due to severe degeneration and displacement of tubular elements, vacuoles appeared in the tubules. Interstitial cell size and number was reduced (Plate 16.3–C, D, E).

Table 16.3: Effect of Cadmium Chloride Treatment on Testes Histomorphology of the Fish *C. orientalis* (High Concentration 15 mg/l)

Experimental Period (h)	*Testes Diam. (μ)*	*Spermatogonia*		*Spermatocytes*		*Spermatids*	*Spermatozoa*
		Primary	*Secondary*	*Primary*	*Secondary*		
Control	798.186±68.193	6.304±1.513	7.843±1.008	48.521±3.697	35.216±2.950	191.392±14.082	166.314±15.802
24–48	712.509*±66.330	5.534*±0.927	6.386*±0.898	43.611*±2.416	26.374*±2.401	157.670*±18.702	155.646*±14.150
	(–10.733)	(–12.214)	(–18.577)	(–10.119)	(–26.232)	(–17.572)	(–6.414)
72–96	700.514*±65.776	5.441*±0.864	6.201*±0.872	43.116*±2.280	25.883*±2.328	153.349*±19.177	154.083*±13.933
	(–12.236)	(–13.689)	(–20.935)	(–11.139)	(–26.502)	(–19.877)	(–7.015)
168–360	679.454*±65.557	5.290*±0.756	5.979*±0.847	42.427*±2.082	23.516**±2.210	147.619*±19.845	152.421*±13.664
	(–14.875)	(–16.085)	(–23.766)	(–12.559)	(–33.223)	(–22.870)	(–8.353)
360–720	651.211±65.211	5.083*±0.597	5.707*±0.793	41.364*±1.802	20.410***±2.044	140.132*±20.725	149.324*±13.324
	(–18.413)	(–19.368)	(–27.234)	(–14.750)	(–42.043)	(–26.782)	(–10.215)

P values: *: < 0.1; **: < 0.01; ***: <0.001.

( ): Parenthesis figures are percentage change.

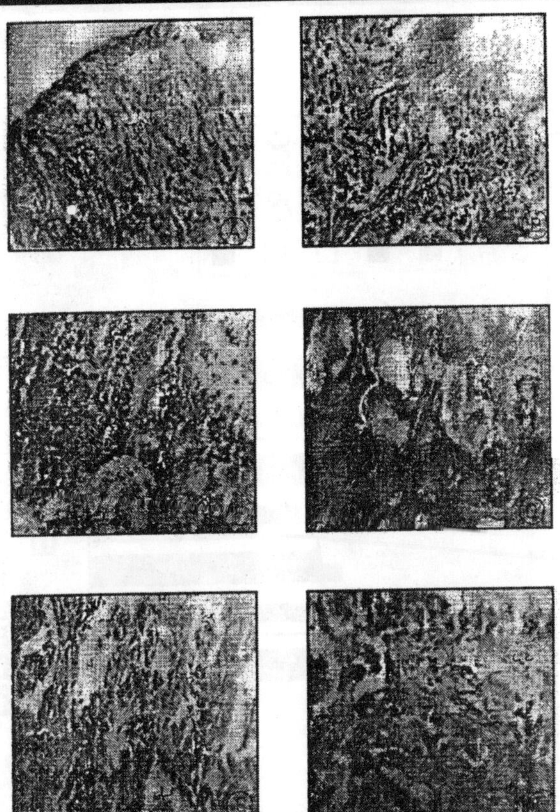

Plate 16.3: Histological Changes in Testes of *Channa orientalis* Exposed of High Concentration of Cadmium Chloride

(A) T. S. of control *C. orientalis* thyroid; (B) T.S. of cadmium chloride exposed *C. orientalis* thyroid during early hours (24–48h.); (C) T.S. of cadmium chloride exposed *C. orientalis* thyroid during middle hours (72–96h); (D) T.S. of cadmium chloride exposed *C. orientalis* thyroid during middle hours (168h); (E) T.S. of cadmium chloride exposed *C. orientalis* thyroid during late hours (360h); (F) T.S. of cadmium chloride exposed *C. orientalis* thyroid during late hours (720h).

PSC: Primary spermatocytes; SSC: Secondary spermatocytes; PSG: Primary spermatogonia; SSG: Secondary spermatogonia; ST: Spermatids; SZ: Spermatozoa; LC: Leydig cells; Nu: Nucleus; V: Vacuoles.

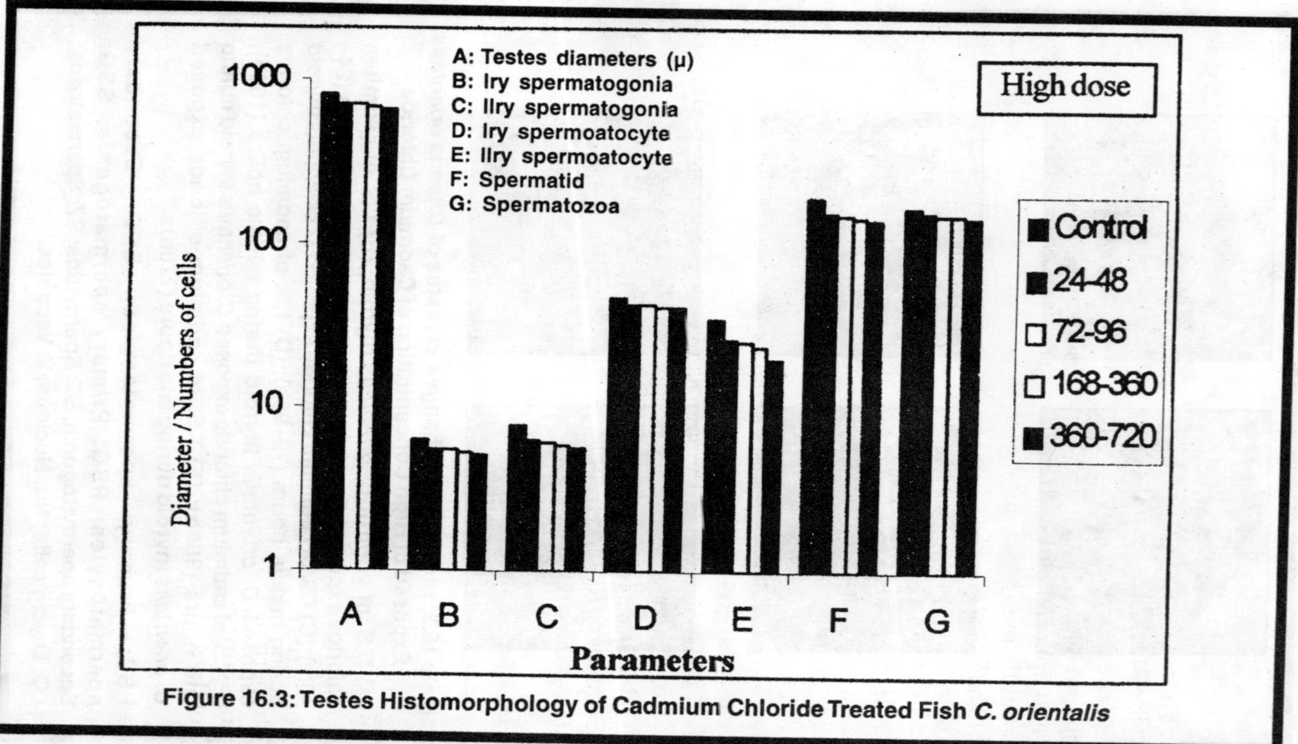

Figure 16.3: Testes Histomorphology of Cadmium Chloride Treated Fish *C. orientalis*

Histological Changes at Low Concentration (5 mg/l) Exposure (Plate 16.4–A, B, C, D, E, F; Table 16.4; Figure 16.4)

The diameter of testes was reduced form control value of 798.003µ to 777.363 µ/24–48 h; 769.250µ 72–96 h; 754.324 µ 68–360 h; 733.663 µ 360–720 h. Seminiferous tubules of central region showed degeneration. Affected tubules showed reduction in cell population. Some of the spermatogonia, spermatocytes, spermatids and spermatozoa showed degeneration. Primary spermatogonia slightly reduced in number from 6.310 (Control value) to 5.936/24–48 h : 5.886/72–96 h; 5.782/168–360 h; 5.669/360–720 h. Depletion in number of secondary spermatogonia were also recorded, (Table 16.4) from control value of 7.850 to 6.472 upto 360–720 hours of treatment. Respective reduction in number of primary spermatogonia and secondary spermatogonia were noted, from (control value) 48.538 to 43.177 (360–720 hours exposure) and 35.206 (control value) to 28.934 (360–720 hours exposure) (Table 16.4). Spermatids and spermatogonia were; also significantly reduced in number. The values were in percent change for spermatids were –8.378, –9.398 –11.064, –12.910 and for spermatozoa were–2.615, –3.115, –3.860, –5.426 at 24–48, 72–96, 168–360, 360–720 hours exposure period (Table 16.4). Leydig cells showed progressive changes followed by reduction in number and size due to significant reduction size of testes the testicular capsules appeared very thin.

Discussion

Thyroid Gland

In *Channa orientalis* after cadmium chloride treatment thyroid gland showed, marked stimulatory effects *i.e.* hypertrophy and hyperplasia in the follicular cells. This observation can be correlated with the fact that pituitary–thyrotrophic cells or cyanophils type–1 of proximal pars distalis showed notable hypertrophy and hyperplasia. According to Bhattacharya *et al.* (1989) ammonia, mercuric chloride, cadmium chloride influences in the functioning of thyroid gland in the fish *Channa punctatus*.

In the present study on *C. orientalis* the hypertrophy, hyperplasia and degranulation in cyanophils–I of proximal pars distalis or thyrotrophs was witnessed while hypertrophy but less secretion in the thyroid follicles suggested that cadmium chloride affects thyroid gland via pituitary.

Table 16.4: Effect of Cadmium Chloride Treatment on Testes Histomorphology of the Fish *C. orientalis* (Low Concentration 5 mg/l)

Experimental Period (h)	*Testes Diam. (μ)*	*Spermatogonia*		*Spermatocytes*		*Spermatids*	*Spermatozoa*
		Primary	*Secondary*	*Primary*	*Secondary*		
Control	798.003±68.206	6.310±1.521	7.850±1.009	48.538±3.682	35.206±2.945	191.376±14.073	166.366±15.810
24–48	777.363NS±66.853	5.936NS±1.350	7.049*±0.970	44.284NS±3.202	31.679*±2.814	175.342*±15.885	161.985NS±15.021
	(–2.586)	(–5.927)	(–10.203)	(–8.764)	(–10.018)	(–8.378)	(–2.615)
72–96	769.250*±66.762	5.886NS±1.306	6.902*±0.961	44.392*±3.113	31.036*±2.776	173.390*±16.223	161.153NS±14.907
	(–3.603)	(–6.719)	(–12.076)	(–8.541)	(–11.844)	(–9.398)	(–3.115)
168–360	754.324*±66.581	5.782*±1.227	6.686*±0.940	43.859*±2.994	30.575*±2.699	170.201*±16.538	159.951*±14.744
	(–5.473)	(–8.367)	(–14.828)	(–9.639)	(–13.154)	(–11.064)	(–3.860)
360–720	733.653*±66.281	5.669*±1.114	6.472*±0.910	43.177*±2.795	28.934*±2.596	166.668*±17.222	157.310*±14.409
	(–8.062)	(–10.158)	(–17.554)	(–11.044)	(–17.815)	(–12.910)	(–5.426)

P values: *: < 0.1; **: < 0.01; ***: <0.001.

Non Significant, ( ): Parenthesis figures are percentage change.

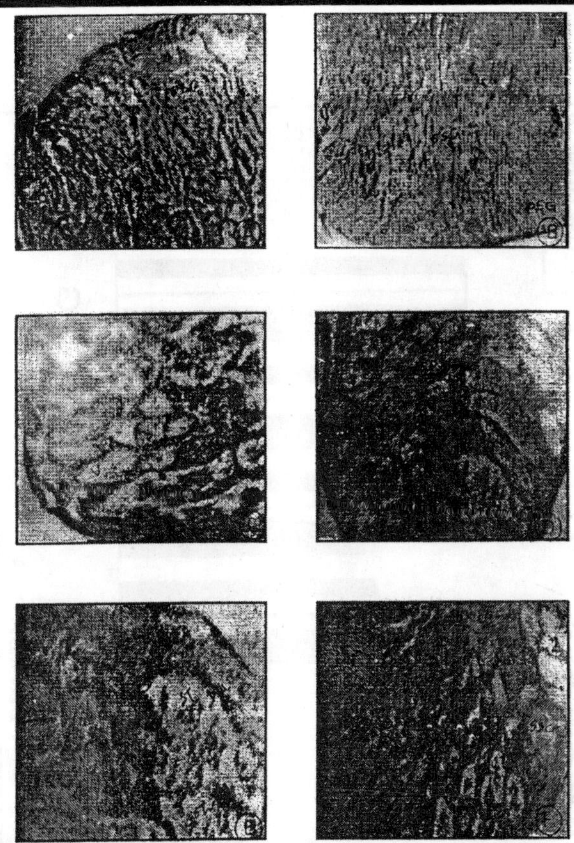

Plate 16.4: Histological Changes in Testes of *Channa orientalis* Exposed Low Concentration of Cadmium Chloride

(A) T. S. of control *C. orientalis* thyroid; (B) T.S. of cadmium chloride exposed *C. orientalis* thyroid during early hours (24–48h.); (C) T.S. of cadmium chloride exposed *C. orientalis* thyroid during middle hours (72–96h); (D) T.S. of cadmium chloride exposed *C. orientalis* thyroid during middle hours (168h); (E) T.S. of cadmium chloride exposed *C. orientalis* thyroid during late hours (360h); (F) T.S. of cadmium chloride exposed *C. orientalis* thyroid during late hours (720h).

PSC: Primary spermatocytes; SSC: Secondary spermatocytes; PSG: Primary spermatogonia; SSG: Secondary spermatogonia; ST: Spermatids; SZ: Spermatozoa; LC: Leydig cells; Nu: Nucleus; V: Vacuoles.

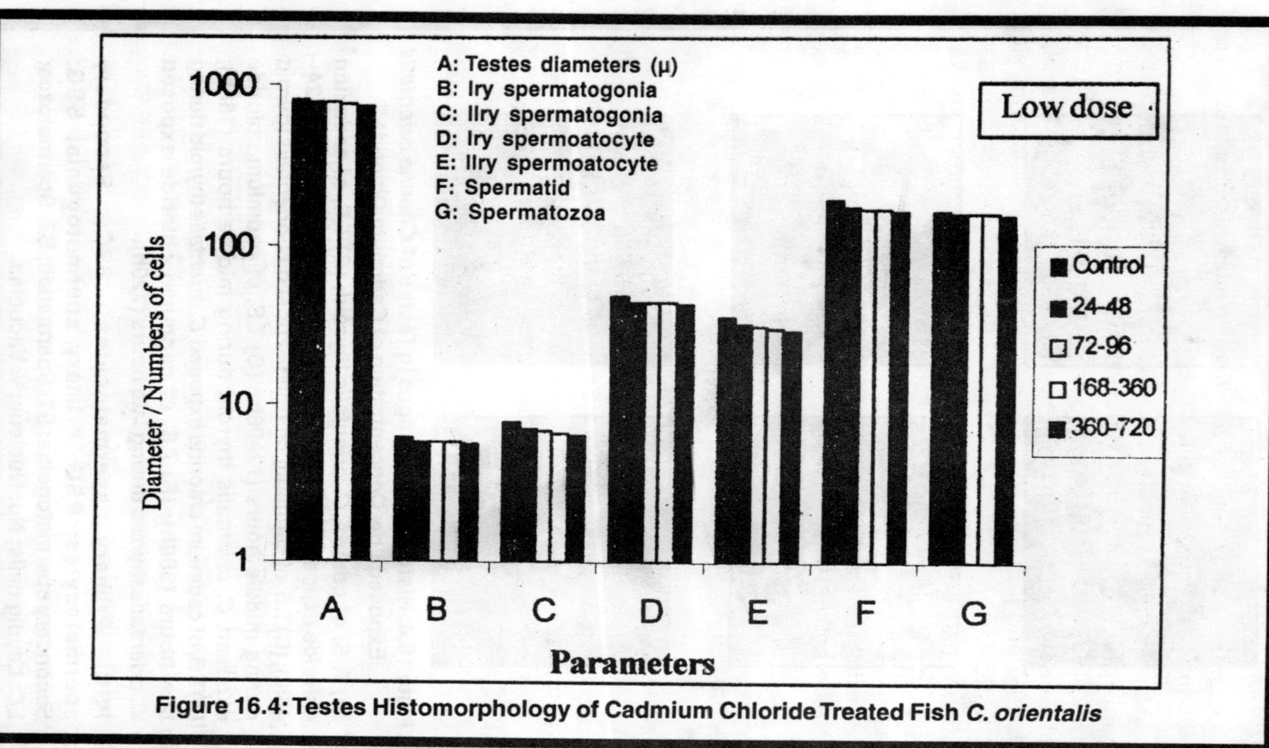

Figure 16.4: Testes Histomorphology of Cadmium Chloride Treated Fish *C. orientalis*

Gonads

Cadmium chloride salt is very toxic to man, other mammalian species but in fishes is degenerative.

Degenerative changes have been recorded in rat testes after the administration of cadmium chloride (Van Campoenhout, 1956; Parizek, 1957 and 1960; Kar *et al.*, 1958 Gunn *et al.*, 1961; Allanson and Deansely, 1962; Mason *et al.*, 1964; Kar and Kamboj, 1965; Gupta *et al.*, 1967 and Johnson, 1967). The inhibition of spermatogenesis and testicular androgen secretion following cadmium chloride treatment had been demonstrated in other mammalian species (Chiquoine, 1963; Clegg and Carr, 1966; Favino *et al.*, 1966; Mackawa *et al.*, 1966; Gunn *et al.*, 1970). It was not known whether these changes in the testes are caused by the direct effect of cadmium chloride on testes or whether it is mediated through the pituitary gland.

According to some investigators vascular injury to spermatic artery is responsible for inhibition of spermatogenesis and regression in leydig cells (Kar and Das, 1960, 1962; Parizek, 1960; Gunn *et al.*, 1963; Cameron and Fostar, 1963; Johnson and Gomes 1970; Setchell and Waites, 1970; Kretster, 1974; Johnson, 1977, Mukharjee, *et al.*, 1984.) Its also reported that cadmium causes specific injury to the internal spermatic artery, its testicular and epididymal branches and pampini form plexus. It was suggested that the break down of testicular vascular system led to an ischemic necrosis of seminiferous tubules and intertubular elements.

However Emmens (1940), Kar *et al.* (1959, 1960 and 1961); Girod (1964 and 1965); Girod and Dubois (1976) and Shrikhande (1983) reported that the cadmium chloride causes direct stimulation of pituitary gonadotrophs.

Perusal of literature indicates that the mechanism of action of this compound is not fully understand.

In the present study cadmium chloride exposed fish *Channa orientalis* showed retardation of gonadal maturation. Cadmium chloride brought about retardation in testicular activities of *Channa orientalis* and caused acute degeneration of testes. The seminiferous tubules, lost their shape. Most of spermatogenic cells became atrophied. The spermatogonia, primary and secondary spermatocytes lost their characteristics and greatly reduced in number. The interstitial cells showed atrophied nature. In these cells cytoplasm was not distinct. Ducts showed more prominent changes.

The fibrous tissue was found to be developed in the walls and the epithelial cells lining these walls which some time were seen atrophied. The testes showed reduction in size.

References

Allanson, M. and R. Deanesly (1962). Observations on cadmium damage and repair in rat testes and the effects on the pituitary gonadotrophs. *J. Endocrinol.*, 24: 453.

Berliner, A.F. and P. Jones-Wittar (1975). Early effects of lethal cadmium dose on Gerbil testes. *Biol. Reprod.*, 13: 240–247.

Bhattacharya, T., S. Bhattacharya, A.K. Roy and S. Dey (1989). Influence of industrial pollutants on thyroid function in *Channa punctatus* (Bloch). *Indian J. Exp. Biol.*, 77: 65–68.

Cameron, E. and C.L. Foster (1963). Observations of the histological effects of sublethal doses of the cadmium chloride in the rabbit. The effects on the testes. *J. Anat.*, 97: 269–280.

Chiquoine, A.D. (1963). Studies on the testicular necrosis induced by cadmium salt. *Ant. Record*, 145: 216.

Chiquoine, A.D. (1964). Observations on the early events of cadmium necrosis of the testes. *Anat. Record*, 149: 23–28.

Clegg, E.J. and Carr J. (1966). Increased vascular permeability in the reproductive organs of cadmium treated male rats. *J. Anat.*, 100: 696–697.

Dixit, V.P. and N.K. Lohia (1974). *Acta. Anatomica*, 89(2): 240–250.

Emmens, C.W. (1940). The production of ovulation in the rabbit by the intravenous injection of salt of copper cadmium. *J. Endocrinal.*, 2: 63.

Favino, A., A.H. Baillie and K. Griffets (1966). Androgen syntheses by the testes and adrenal gland of rat poisoned with cadmium chloride. *J. Endocrinal.*, 35: 185.

Girod, C. (1964). Etude des cellules gonadotrophs an to hypophysaires du singe *Maoacus irus*. F. cus; apre's administration de chloride de cadmium *Cr. Soc. Biol.*, 158: 948.

Girod, C. (1965). Influence du chlorure de cadmium sur l' axe hypophysotesticulaire chezle singe *Macacus irus. F. Cuv. Soc. Biol.*, 158: 2113.

*Girod, C. (1966). *Bull Ass. Anat.*, 132: 451–459.

*Girod, C. and A. Chauvineau (1965). *C.R. Soc. Biol.* (Paris), 158: 2113.

Girod, C. and M.P. Dubois (1974). Influence d' ure injection de chlorure de cadmium sur l' axe hypophysotesticular du hamster dore. (*Mesocricetus auratus* Waterh). Cr. Se onces. *Soc. Biol. Sec. Fil.*, 168: 970.

Gunn, S.A. and T.G. Gould (1970). Cadmium and other mineral elements. In: *The Testes,* (Eds.) A.D. Johnson, W.R. Gomes and N.L. Van Demai... Academic Press, New York, 3: 377–434

Gunn, S.A., I.C. Gould and W.A.D. Anderson (1961). Zinc protection against cadmium injury to rat testes. *Arch. Path.* (Chicago), 71: 274–281.

Gunn, S.A., T.C. Gould and W.A. Anderson (1963). The selective injurious response to testicular and epididymal blood veseels to cadmium and its prevention by Zinc. *Am. J. Path.*, 42: 685–702.

Gunn, S.A., T.C. Gould and W.A.D. Anderson (1970). Cadmium and other mineral elements. In: *The Testes,* (Eds.) A.D. Johnson, W.R. Gomes and N.L. Van Demark. Academic Press, New York, 3: 377–481.

*Gupta, R.K., G. W. Berns and F.R. Skelton (1977). *Amer. J. Pathol.*, 51: 191.

Johnson, A.D. (1977). *The Testes,* Vol. IV. Academic Press, New York and London.

Johnson, A.D. and W.R. Gomes (1970). Cadmium and other mineral elements. In: *The Testes.* Academic Press, New York, 3: 377–481.

Kar, A.B. and R.P. Das (1960). Testicular changes in rats after treatment with Cadmium chloride. *Acta. Biol. Ger.*, 5: 153–173.

Kar, A.B. and R.P. Das (1962). Sterilization of males by intratesticular administration of Cadmium Chloride. *Acta. Endocr. Copenh.*, 40: 321–331.

Kar, A.B. and V.P. Kamboj (1965). Cadmium damage of the rat testes and its prevention. *Indian J. Exp. Biol.*, 3(1): 45–49.

*Kar, A.B., Karkun, J.N. and R.P. Das (1958). *Indian J. Med. Res.*, 47: 20.

Kar, A.B., R.P. Das and J.N. Karkun (1959). Ovarian changes in prepubertal rats after treatment with cadmium chloride. *Acta Biol. Med. Ger. J.*, 372–399.

Kar, A.B., P.R. Dasgupta, and R.P. Das (1960a). Effect of Cadmium chloride on gonadotrophin content of the pituitary of male and female rats. *J. Sci. Ind. Res.* (India), 19C: 225.

Kar. A.B., Dasgupta and R.P. Das (1961). Effects of low dose of cadmium chloride on the genital organs and fertility of male rats. *J. Sci. Indust. Res.*, 20C: 322–326.

*Kretser, D.M. (1974). *Contraception*, 9(6): 561–600.

Maekawa, K.Y., K. Tsunenari and V. Kurematsu (1966). Role of increased vascular permeability in cadmium injury of the testes. *Acta. Anat. Nippon*, 41: 327–333.

Mason, K.E., J.A. Brown, A.D. Young and R.R. Westbit (1964). Cadmium induced injury of rat testes. *Anat. Rec.*, 149: 155.

Meek, E.S. (1959). Cellular changes induced by cadmium in mouse testes and liver. *British J. Expl. Pathol.*, 40: 503–506.

Mukherjee, M.M., N. Ganguly and S.R. Dasgupta (1984). Short and long-term effects of Cadmium chloride on the testes of rats. *J. Anant. Soc.*, India, 33(2): 107–115.

Perizek. J. (1960). Sterilisation of male rats by Cadmium salt. *J. Reprod. Fertil.*, 1: 291–309.

Perizik, J. (1957). The destructive effects of Cadmium ion on testicular tissue and is prevention by Zinc. *J. Endocrinol.*, 15: 56–63.

Ramaswamy, L.S. and D.K. Kual (1966). The effect of Cadmium and selenium of the testes of the dessert gerbil *Mariones hurrianae Jerdon* Four. *Roy. Micro. Soc. London*, 85(3): 297–307.

*Setchell, B.P. and G.M.H. Waiter (1970). *J. Endocrinol.*, 47: 81.

Sharma, G.P., M.L. Sareen and V.K. Monga (1980). Effect of Cadmium chloride on the testes of the guinea-pig. *Cavia cobaya. Proc. Nat. Acad. Sci., India*, 50(B): 111.

Shrikhande, D.J. (1983). An experimental analysis of certain pituitary–adrenal gonadal–relationship in India fruit bat–Rousettus leschenaulti. *Ph.D. Thesis.* Nagpur University, Nagpur.

*Singwi, M.S. and S.B. Lal (1980). *Indian J. Exp. Biol.*, 18: 1398–1401.

*Van Campoenhout, E. (1956). *Archiv. Biol. (Liege)*, 67: 499–512.

*: Not seen in original.

Chapter 17

Heavy Metal Concentrations in the Edible Crab *Scylla serrata* in the Malancha Region of Indian Sundarbans

Kakoli Banerjee, Abhijit Mitra, Rajib Chakraborty, Anumita Das and Debarati Mukherjee

ABSTRACT

Concentrations of Zn and Cu were estimated in the muscle of the edible crab Scylla serrata inhabiting the Malancha region of Indian Sundarbans in different seasons. Highest concentrations of these metals were detected during the pre-monsoon. In the monsoon months of low aquatic salinity, pH and high dilution factor, the metal level in the body tissue of the detritivorous crabs decreased, which may be as per the biologically available heavy metals in the surface sediments.

Keywords: *Scylla serrata, Edible crab, Metals, Salinity, pH, Detritivorous.*

Introduction

Production of heavy metals and their discharge into the aquatic environment have increased greatly in recent years. This results in the deposition of heavy metals on the underlying sediment bed.

Heavy metals find their way into the marine and estuarine environment from various sources:

1. Pesticides and insecticides from agricultural fields carried by streams, rivers and estuaries.
2. Industrial effluents of diverse types discharged directly or along with domestic sewage.
3. Radioactive wastes discharged from nuclear power stations.
4. Waste heat discharged from thermal power stations.
5. Solid wastes and military wastes.
6. Debris from ocean bed exploration for oil and minerals.

A fairly good amount of these discharged metals accumulate in the body tissues of benthic organisms and hence they are often considered as potential indicators of pollution. The availability and magnitude of heavy metals in the body tissue is dependent on several factors such as nature of chemical species (Waldichuk *et al.*, 1974), salinity (Bryan and Uysal, 1978), temperature (Sastry and Moller, 1981), pH (Ramamoorthy, 1988) etc.

It is again recommended that the damage of estuarine ecosystem is mainly a function of biologically available metallic fraction rather than the total amount of metal present in water or in sediment (Chester and Stoner, 1975; Chester and Voutsinou, 1981).

Based on these informations, an attempt has been made in the present programme to determine the seasonal pattern of metal accumulation by the edible crab *Scylla serrata* collected from Malancha region of North 24 Parganas, Sundarbans, West Bengal, India. The study was conducted for a stretch of two years during 2002 and 2003.

Materials and Methods

The estuarine system of West Bengal (India) is one of the most biologically productive, taxonomically diverse and aesthetically celebrated ecotone of the country and is supporting the world's largest mangrove block, the Sundarbans. This system is a unique nursery ground for a wide variety of fin fish and shell fish juveniles. Based on the culture period of crabs in the study area, crabs were collected monthly during March, 2002 to December, 2002 and March, 2003 to December, 2003 from the Malancha region of Indian

Sundarbans. Precaution was taken to maintain a standard size during crab collection so as to avoid any error due to size difference.

Soft parts were carefully removed after quick rinse with double distilled water, and the body parts were oven dried at 110°C. The metals (Zn and Cu) were estimated in the dried samples after digestion with conc. H_2SO_4 and conc. $HClO_4$ in the ratio of 3 : 1. The resulting solution was digested and made upto a constant volume with 0.05 N HNO_3. The analytical blanks were prepared and treated with the same reagents. The digested samples were analysed for Zn and Cu through a Perkin Elmer Atomic Absorption Spectrophotometer (AAS Model 3030) equipped with a HGA-500 graphite furnace atomizer and a deuterium background corrector.

Results and Discussion

Concentration of metals in the body tissues varied markedly, showing a definite pattern with season. In general, high concentrations were observed during the pre-monsoon months (March to June), the period characterized by highest salinity and pH of the overlying aquatic phase. This condition of the water favours the process of precipitation of heavy metals on the sediment as observed by earlier workers (Mitra *et al.*, 1992; Trivedi *et al.*, 1995; Choudhury *et al.*, 1994; Gupta *et al.*, 1994). As the crabs are detritivores, therefore the body burden of the selected heavy metals also increased in the pre-monsoon season (Tables 17.1 and 17.2; and Figures 17.1–17.4). *Scylla serrata* being an important edible species fetches high price in the international market; hence such study is of utmost importance from the quality point of view.

Table 17.1: Monthly Variation of Zn and Cu (in µg/gm dry wt.) in the Body Tissue of the Edible Crab During 2002

Month	*Zn Conc. (µg/gm)*	*Cu Conc. (µg/gm)*
March	520.56	225.35
April	609.83	232.81
May	616.84	281.66
June	648.29	341.04
July	605.02	146.31
August	398.42	105.44
September	311.37	90.00
October	401.56	108.13
November	517.24	137.40
December	528.88	117.04

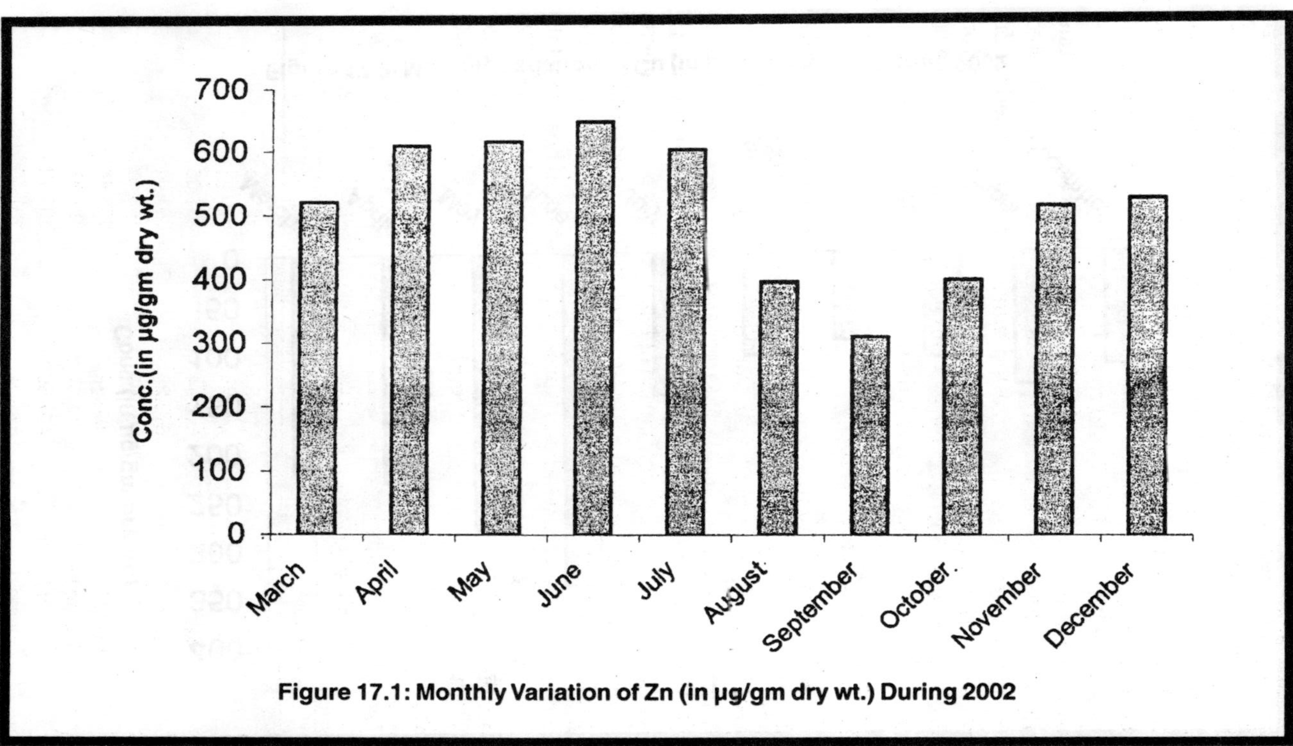

Figure 17.1: Monthly Variation of Zn (in µg/gm dry wt.) During 2002

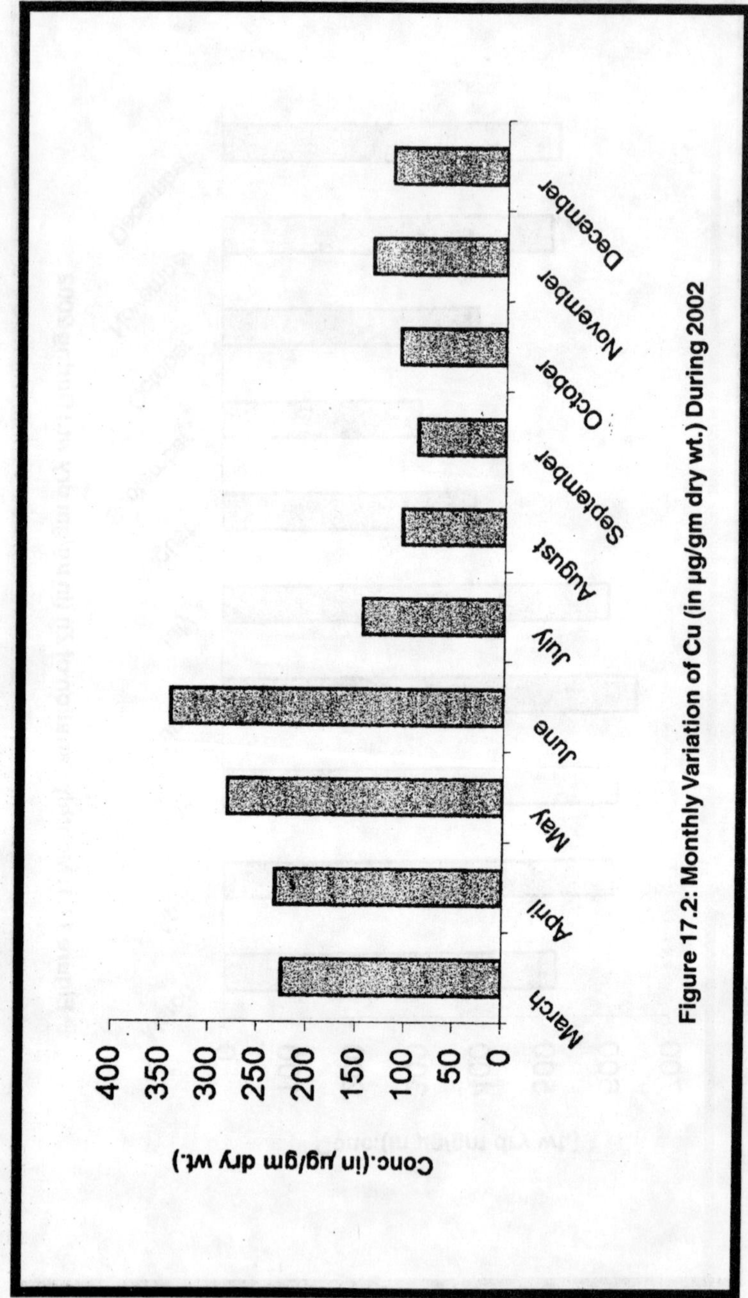

Figure 17.2: Monthly Variation of Cu (in µg/gm dry wt.) During 2002

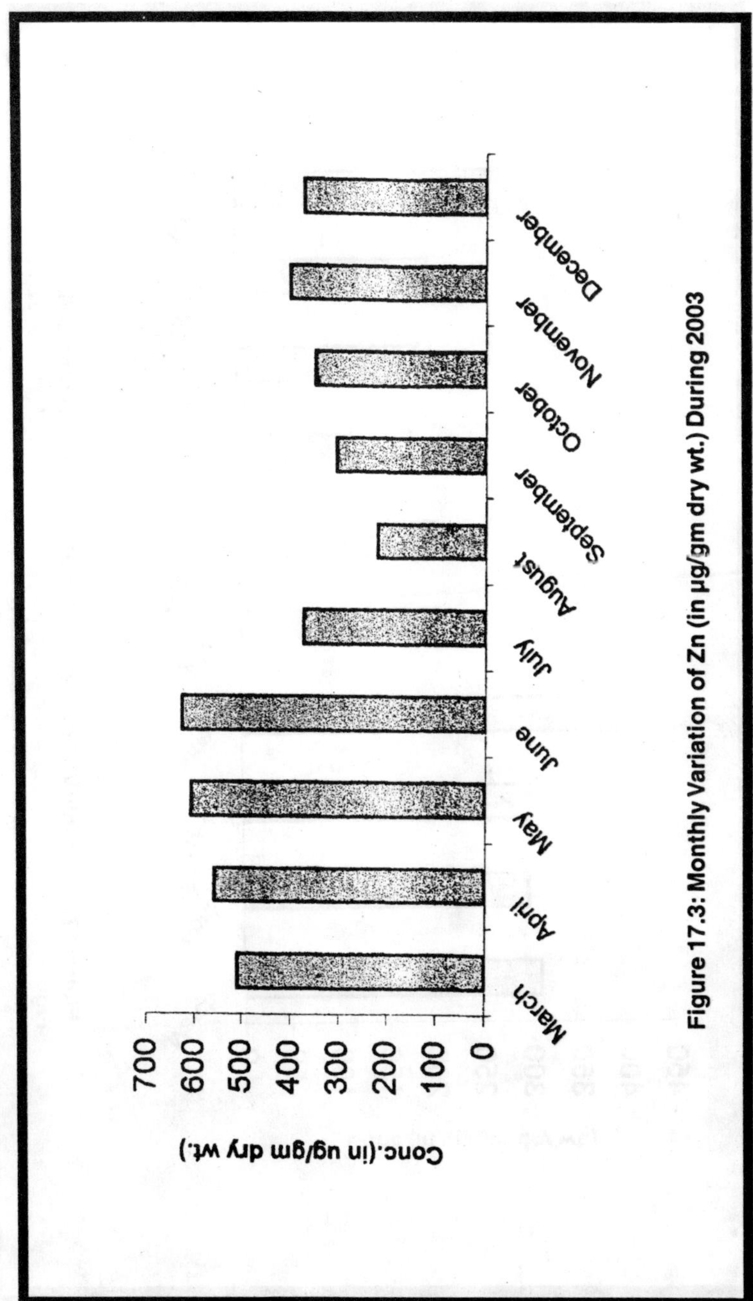

Figure 17.3: Monthly Variation of Zn (in µg/gm dry wt.) During 2003

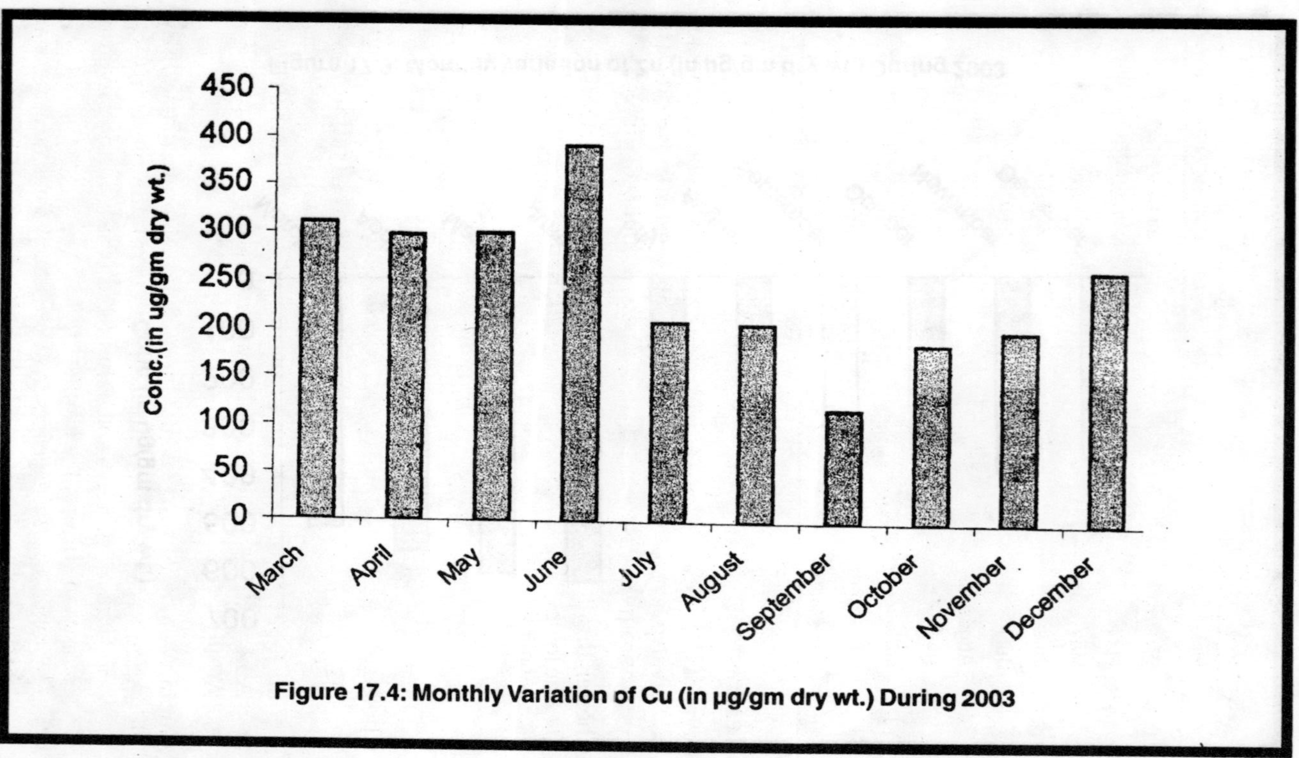

Figure 17.4: Monthly Variation of Cu (in µg/gm dry wt.) During 2003

Table 17.2: Monthly Variation of Zn and Cu (in µg/gm dry wt.) in the Body Tissue of the Edible Crab During 2003

Month	*Zn Conc. (µg/gm)*	*Cu Conc. (µg/gm)*
March	509.45	311.29
April	559.58	298.96
May	608.83	301.65
June	628.05	393.46
July	373.36	210.30
August	220.42	208.59
September	306.73	119.02
October	351.02	188.39
November	402.52	202.54
December	373.43	266.28

References

Bryan, G.W. and H. Uysal (1978). Heavy metals in the burrowing bivalve *Serobicularia plana* from the Tamar estuary in relation to environmental levels. *J. Mar. Biol. Ass.* (UK), 58: 89–108.

Chester, R. and J.H. Stoner (1976). Trace elements in sediments from the lower Severn estuary and Bristol Channel. *Mar. Pollut. Bull.*, 5: 92–95.

Chester, R. and F. Voutsinou (1981). The initial assessment of trace metal pollution in coastal sediment of Hooghly estuary. *Environ. Ecol.*, 7: 427–430.

Choudhury, Abhijit; Mitra, Abhijit; Trivedi, Subrata; Gupta, Ananda and Amalesh Choudhury (1994). Phosphate and nitrate status in the east coast of the Indian subcontinent. *Proceedings of Seminar on Our Environment: Its Challenges to Development Projects.* American Society of Civil Engineers–India Section. p. AC-1–AC-9.

Gupta, Ananda; Mitra, Abhijit, Mitra; Trivedi, Subrata; Choudhury, Abhijit and Amalesh Choudhury (1994). Simple Statistical approach to underst and fate of pollutants in the Hooghly estuary in different seasons. *Proceedings of Seminar on Our Environment: Its Challenges to Development Projects.* American Society of Civil Engineers–India Section. p. AT-1–AT-6.

Mitra, Abhijit; Choudhury, Amalesh and Yusuf Ali Zamaddar (1992). Effects of heavy metals on Benthic Molluscan Communities in Hooghly estuary. *Proc. Zool. Soc., Kolkata*, 45: 481–496.

Ramamoorthy, S. (1988). Effect of pH on speciation and toxicity of aluminium to rainbow trout (*Salmo qairdneri*). *Can. J. Fish. Aquat. Sci.*, 45: 634–642.

Sastry, A.N. and Don C. Moller (1981). Application of Biochemical and Physiological Responses to Water Quality Monitoring. In: *Biological Monitoring of Marine Pollutants*, (Eds.) J. Vernberg, A. Calabrese, F.P. Thurberg and W.B. Verberg. Academic Press. New York, p. 265–294.

Trivedi, Subrata; Mitra, Abhijit; Bag, Manigrib; Ghosh, Indranil and Amalesh Choudhury (1995). Heavy metal concentration in Mud Skipper *Boleophthalmus boddaerti* of Nayachar Island, India. *Indian J. Env. Hlth., NEERI.* 37(2): 120–122.

Waldichuk, M. (1974). Some biological concerns in heavy metals pollution. In: *Pollution and Physiology of Marine Organisms.* Academic Press, New York, p. 1–58.

Chapter 18

Histological Alterations in Tadpoles of Bufo Melanostictus, Exposed to a Sublethal Concentration of Chromium

D. Anusuya, D.J. Prakash & I. Christy

ABSTRACT

Effect of sublethal concentration of heavy metal chromium (0.2 ppm) on tadpoles of Bufo melanostictus was studied. Results reveal extensive histological abnormalities in the internal gills, intestine, liver and eye of tadpoles exposed to chromium, in contrast to that of control tadpoles, showing the severity of chromium toxicity.

Keywords: *Chromium, Tadpole, Histological alterations, Necrosis, Sublethal, Toxicity.*

Introduction

Heavy metals, are more hazardous pollutants of ecosystems with deleterious effects on aquatic biosystems. Rapid industrialization, urbanization and other developmental activities have led to deterioration of the environment (Salunk *et al.*, 1982). Such heavy metal toxicants have been shown to enter and accumulate in aquatic fauna and flora causing heavy modification

(Berman and Lal, 1994). Earlier reports reveal that they may disturb cellular functions, altering vital physiological and biochemical mechanisms of animals (Larsen *et al.*, 1976; Shastry and Rao 1981; Muthukrishnan *et al.*, 1986; and Radhakrishniah *et al.*, 1991) because of their inherent ability to form complexes with ligands containing sulphur, nitrogen and oxygen as electron donors (Valee and Ulmer, 1972). The deleterious effects caused by the toxicity of chromium, have been reported and reviewed (Mertz, 1969; Binanchi and Lewis 1986; Anna M. Fan and Harding Barlow, 1987). Extensive information is available on effects of chromium of Ichthyofauna (Venugopal and Reddy, 1961; Khaviraj, 1984 and Srivatsava and Mauria, 1991). There is a paucity of information on such effects in amphibian population excepting few reports (Mckee and Wolfe 1963; Greenhouse 1976; Schowing and Boverio 1975). Further it is also essential to evaluate toxic effects of chromium or any heavy metal on amphibians, since frogs and toads form very important members of food chain in aquatic as well as terrestrial ecosystem (Leuven *et al.*, 1986). It has also been established that amphibian embryos and larvae could be used as sensitive assay system to study and evaluate toxic and teratogenic effects of aquatic pollutants (Birge *et al.*, 1975 and Greenhouse, 1976). Therefore, the present investigation aims to evaluate the histopathological abnormalities caused in some selected tissues of tadpoles of Bufo melanostictus due to chromium toxicity.

Material and Methods

Tadpoles of Bufo melanostictus, ranging from 1.2 to 1.5 cms were collected from near by water bodies and acclimated in bore well water which had the characteristics such as temperature 28°C, pH 8.2, dissolved oxygen 5.2 ± 0.4, for two days before commencing the experiments.

Suitable stock solution of technical grade Potassium dichromate (28.28 gms/litre) was prepared from which required number serial dilutions were made and static bio-assays were conducted as per the procedure (APHA 1974). LC_{50} for 96 hr value was determined following the procedure of Reish and Cshida (1987) and $1/10^{th}$ of LC_{50} value (0.2 ppm) was taken as sublethal concentration as suggested by Kirubakaran (1998) and the same was used in the present study. The experimental animals were kept in chromium containing water (0.2 ppm) while the control tadpoles were kept in bore well water. During the experimental period, the medium was changed once a day and the tadpoles were fed with freshly boiled spinach leaves. The total period of experimentation lasted for 20 days.

At the end of experimental period, tadpoles were sacrificed and fixed in 5 per cent formaldehyde and processed for histological studies. Serial sections of 7 μm thick were taken and stained with haematoxylin and eosin. The sections were identified and photographed.

Results and Discussion

Histological approach is the most valuable tool for assembling the action of toxicant at tissue level providing data concerning tissue damage (Sprague 1971). Visible histopathological abnormalities caused due to toxicity of heavy, metals in animals have been reported earlier (Srivatsava *et al.*, 1982; Ramesh Kumar *et al.*, 1988 and Srivatsava and Maurya 1991). The present study also reveals me extensive damage in the internal gill architecture of treated tadpoles compared to gills of controls.

As observed in fish by earlier workers, there is a clear loss of receptor epithelial layer, showing erosion of epithelial cells with necrosis of tissues in the chromium treated tadpoles in contrast to intact gill structure of control tadpoles. Intestine is the vital organ for digestion and any food contaminated with a toxicant is known to cause damage to the intestinal epithelium (Amita Moitra and Roshan Lal, 1989).

In the present study, chromium exposed tadpoles showed visible alterations in intestinal epithelium. In the control, intestinal epithelium was intact with more granulated and deeply stained cells where as chromium treated tadpoles exhibited thickened, broken and lightly stained mucosal layer with dispersed cells.

Liver is an organ to metabolize and eliminate any toxicant or drug and any histopathological changes observed in liver due to impact of toxicant can be used as a reliable index to evaluate structural and functional damage of target tissues (Shastry and Sharma, 1978) and the extent of its damage is related to the mode of action of toxicant, its accumulation and persistence (Ravi Kumar 1994). In the present study, the liver of control tadpoles showed in intact histoarchitecture with deeply stained nuclei and granulated cytoplasm. In contrast, liver of treated tadpoles revealed enlarged cells, lightly stained cytoplasm, extensive vacuolization and shifting of nuclei to the periphery of some cells. These findings agree with the earlier reports (Bakthavatchalu, 1984 and Narain and Singh, 1991) which showed characteristic abnormalities like liver chord disarrangement, nuclear pycnosis and degeneration of cytoplasm in fishes exposed to various pesticides.

The eyes are in direct contact with the external medium and therefore its exposure to toxicant medium in the treated tadpoles, has resulted in erosion of sclerotic coat, chord layer, and a collapsed lens vesicle in contrast to compact nature of eye structure with intact lens, retinal layer, choroid layer and sclerotic coat of control tadpoles.

Thus the present observations clearly indicate that even a sublethal concentration of heavy metal chromium could result in severe histological impairments of various tissues in tadpoles exposed for a considerable length of time.

References

Anna, M. Fan and I. Harding Barlow (1987). In: *Advances in Modern Environmental Toxicology.* (Eds.) M.A. Mehlmon Princeton. Scientific Publishing Co. Inc., p. 87–125.

APHA (1974). *Standard Methods for the Examination of Water and Wastewater*. APHA, AWWA and WPCF, Washington, p. 1134.

Bakthavatchalu, R. (1984). Histomorphology of liver, kidney and intestine of the fish. Anabus testudian exposed to Furan. *Environ. Ecol.*, 2: 243–8.

Berman, S.C. and M.M. Lal (1994). Accumulation of heavy metals (Zn, Cu, Cd, and Pb) in soil and cultivated vegetables and weeds grown in industrially polluted fields. *J. Environ Biol.*, 15(2) 107–115.

Bianchi and A.G. Lewis (1986). Mechanism of chromium genotoxicity. *Toxicol. Env. Chem.*, 9: 1–25.

Brige, W.J., Westerman, A.G. and J.A. Black (1975). Sensitivity of vertebrate embroys to heavy metals as a criterion for water quality. Phase III. Use of fish and amphibian embroys as bioindicator organisms for evaluation of water quality. Res. Rep. University of Kentuky, Water Resource Inst., U.S.A. No. 91.

Greenhouse (1976B). Evaluation of the teratogenic effects of hydrazine, methylhydrazine and dimethyl hydrazine on embryos of Xenopus laevis, the South African toad. *Teratology*, 13: 167–178.

Khaviraj (1984). The effect of copper on maximum respiration rate and growth rate of puch perca fluviatilis. L., *Water Research.*, 18: 30–144.

Kirubagaran, R. (1989). Endocrine and neuroendocrine physiology of the cat fish clarious batrachus in relation to mercury treatments. *Ph.D. Thesis*, Centre of Advance study in Zoology, Banaras Hindu University.

Larson, A. Bengteson and B.E. Stevenberg (1976). Some hematological and biochemical effects of CCI in fish. *Sem. Ser. Soc. Exp. Biol.*, 2:35–46.

Leuvan, R.S.E.W., Denttar, C., Christian, M.M.C. and W.H.C. Heijligers (1986). Effect of water acidification on the distribution pattern and reproductive success of amphibians. *Experentia*, 42: 493–503.

McKee, J.E. and H.W. Wolf (1963). Water quality Criteria publication of the California state water resources control board, U.S.A. Pub. No. 3: A revised reprint, 1976.

Mertz, W. (1969). Chromium occurrence and function biological systems. *Physiol. Res.*, 49: 163–234.

Moitra, Amita and Roshan Lal (1989). Effect of sublethal doses of Malathion and BHC on the intestine of fish, *Puntius sarana. Env. & Ecol.*, 7 (2): 412–414.

Muthukrishnan, J., Viswarajan, S. and E. Subbulakshmi (1986). Effect of sublethal concentration of Mercuric Chloride on transformation of food by the fish *Cyprinus Carpio. Environ. Ecol.*, 4: 526–532.

Narain, A.S. and B.B. Singh (1991). Histopathalogical Lessions in *Heteropneustes fossils* subjected to acute thiodon toxicant. *Acta Hydrochem. Hydrobiologi*, 19: 235–243.

Radhakrishnaiah, K., Suresh, A., Urmilaa, B. and B. Sivaramakrishnan (1991). Effects of mercury and liquid metbolic profiles in the Organs of *Cyprinus Carpio* (Lin). *J. Mendal*, 8: 125–135.

Rameshkumar, P.B., Vijayalakshmi, S. and C. Rajamanikam (1988). Toxicity effect of Zinc Sulphate on gill in the freshwater fish *mystus vitatts* (Block). Asbst No. 67, National Symposium on Ecotoxicology, Annamalai Nagar.

Ravikumar, P. (1986). Impact of Phosphomidon of histological and biochemical changes in the FW Fish *Oreochromis Mossambicus* (Peters). *M.Phil. Thesis*, Annamalai University.

Reish, L. Donald and Philips Cshida (1987). Manual of methods in aquatic environment research, 247, Part–10: short term static bio assays, FAO, Fisheries Technical Paper, Rome.

Salunk, J., Balogh, V.K.O. and E. Berta (1982). Heavy metals in animals of the lake Belaton. *Water Res.*, 16: 1147–1152.

Schowing and Boverio (1979). Influence teratogene du bichlorure demercure surle development. Embryonairede, amphibian Xenophous laevis, Acta, Embryol expt., 1: 39–52.

Shastry, K.V. and S.K. Sharma (1978). Eldrin toxicity on liver of *Channa Punctatus* (Block). *Ind. J. Exp. Biol.*, 16: 372–373.

Sprague, J.B. (1971). Measurement of pollutant toxicity to fish. III, Sublethal effects and sate concentration. *Water Res. J.*, p. 245–260.

Srivastava, V.M.S. and R.S. Maurya (1991). Effect of Chromium stress on gill and intestine of *Mystus Vittatus* (Block: Scanning E.M. Study). *J. Ecobiol.*, 3(1): 69–71.

Srivastsava, V.M.S., Tripathi, R.S. and A.K. Saxena (1982a). Chromium included histopatholgic changes in some tissues of *Puntius sophore* (Hamilton). *J. Biol. Res.*, 2: 67–68.

Valee, B.L. and D.D. Ulmer (1972). Biochemical effects of Mercury, Cadmium and lead. *Ann. Rev. Biochem.*, 41: 91.

Venugopal, N.B. and S.L. Reddy (1992). Nephrotoxic and hepatotoxic effects of trivalent and hexavalent Chromium in a teleost fish, *Anabas Scandans*, Enzymological and Biochemical changes. *Ecotoxicol. Env. Safety*, 24(3): 287–93.

Chapter 19

Cadmium and Lead Level in a Semi-urban Area and their Poisoning Effect

M.K. Paul and A.K. Misra

ABSTRACT

The present article includes the study of cadmium and lead as heavy metals present in drinking water of a semi-urban area of Lumding Town. Presence of these two metals in some sources were discussed and how these two metals affect in human body and their metabolic function were discussed.

Keywords: *Cadmium, Lead, Metabolic affect.*

Introduction

The heavy metals are used directly or indirectly by the human being and other organisms. When these metals are absorbed by the absorbing living materials and ultimately crosses a limit which are hazardous for them, then these metals are called toxic metals. Cadmium and Lead are categorized in these group. The status of these trace elements–the quantitative concentrations and their affect in human body is described here.

Heavy metal pollution is a matter of concern nowadays and out break of many diseases are due to them. Lumding is a semi-urban railway town of south-east part of Nagaon district of Assam, India. Some drinking water sources are taken as random to study the heavy metals (HM) Cadmium and Lead present in these sources (Table 19.1).

Table 19.1: The Various Sampling Stations and their Sources

Sl.No.	*Nature of Source*	*Sampling Stations*
1.	Dugwell	Haflong Road
2.	Dugwell	Nadirpar
3.	Ringwell	Santipara
4.	Ringwell	Bazar Area (Main Bazar)
5.	Ringwell	New Colony
6.	Ringwell	Buddha Mandir Area
7.	Pond	Haflong Road
8.	Pond	Jarangdisha
9.	Railway Supply Water	Railway supply water
10.	PHE supply water	ASEB Area

Methodology

The samples were collected in pre-cleaned polythene bottles and then rinsed by dil. HNO_3. Then they were analyzed by APHA (1995, 19th Edition). They were analyzed by Atomic Absorption Spectrophotometer by direct Air-Acetylene method (Perkin-Elmer Model 3280). All the experiments were done in various season from May 2002 to April 2004. The obtained results were compared by USPH, WHO, European Standard, ICMR and ISI values with respect to drinking water standard (Table 19.2).

Table 19.2: Parameters for Water Quality Characterization and Standards (Domestic Water Suppliers) Maximum Permissible Limits

Parameters	*USPH Standard (mg/l)*	*WHO Standard (mg/l)*	*European Standard (mg/l)*	*ICMR Standard (mg/l)*	*ISI Standard (mg/l)*
Cd	0.01	0.01	0.05	–	0.01
Pb	0.05	0.1	0.1	0.1	0.01

Table 19.3: All Season Ranges of Cd and Pb in Dugwells, Ponds, Ringwells, Railway Supply Water, PHE Supply Water

Sampling Source (S)	*Cd (mg/l)*	*Pb (mg/l)*
S-1	ND-0.01	ND
S-2	ND-0.01	ND
S-3	ND-0.01	ND
S-4	ND	ND-0.01
S-5	ND	0.01-0.03
S-6	ND-0.02	ND-0.03
S-7	ND-0.02	ND-0.01
S-8	ND	ND
S-9	ND	ND
S-10	ND	ND

Results and Discussion

The various water sources for drinking purposes are dugwells (DW), Ringwells (RW), Ponds (P), River (R), railway supply water (RSW) and Public Health Engineering (PHE) water supply. Ten sources are selected from the Town. The minimum and maximum level of Cd and Pb were determined.

Cadmium

The source S-4, S-5, S-8, S-9, S-10 have the cadmium values below detection limit and specified as not detected (ND) other five sources–minimum levels is not detected (ND) to 0.01 mg/l and not detected (ND) to 0.02 mg/l. The sources S-1, S-2, S-3 are ND to 0.01 mg/l and the sources S-6, S-7 are ND to 0.02 mg/l. The maximum permissible limit of Cadmium by USPH, WHO and ISI limit of 0.01 mg/l and by European standard is 0.05 mg/l. So cadmium levels of all the samples are within the regulating agencies.

Toxicology

Cadmium has no positive role in biological function of human but weak evidence for ultra trace essentially in rats. Cadmium is highly toxic which can cause a cumulative poison in mammals, causes renal dysfunction, hypertension, testicular tumours, gastrointestinal upsets, bronchitis, emphysema, anemia, prostrate cancer, nephritis, itai-itai byo (wrong bone metabolism diseases).

Tobacco byproduct is a potential source of Cd-poisoning. On smoking a cigarette 1.4 μg Cadmium comes to our ecosystem. Smoking 10–20 cigarettes per day can increase the body burden of Cd by 1–4 μg. The various sources of Cd along with Zn commercially as galvanization, brass, dry cells, pigment, soaps, textile, paper, rubber, printing inks, ceramic glazes and fireworks.

Cd^{+2} leads to decalcification in bones and bones become weaken. Cd^{+2} poisoning produces kidney problem, Cd^{+2} replaces the S-atom of SH group in the enzymes Cadmium replaces the metal Cu, Fe, Zn from the body and makes deficiency of essential metal. Cd^{+2} can inhabit several DNA repair enzymes.

Lead

The source S-1, S-2, S-3 and S-8, S-9, S-10 have the Lead below detection limit *i.e.* not detected (ND). Other four sources minimum level of Lead is from not detected to the maximum level 0.01 mg/l and 0.03 mg/l. The source S-4 and S-7 have Lead level from ND to 0.01 mg/l the source S-5 has 0.01 mg/l to 0.03 mg/l and S-6 has ND to 0.03 mg/l of Lead. The maximum permissible limit of Pb by WHO, European standard, ICMR is 0.1 mg/l and by USPH is 0.05 mg/l. So the Lead of all the tested samples are within these regulating agencies.

Toxicology

Biological function of Lead is not known. It is very toxic to most plants, cumulative poison to mammals. Lead enters in the aquatic system through precipitation, Lead dust fall out Leaded petrol, erosion and leaching of soil, waste discharges and run off Lead pluming. The solubility of lead increases with temperature (Subraniam, 1995). It inhabits δ–aminolevulinate dehydrase and thus hemoglobin synthesis is stopped. As a result anemia takes place. Pb^{+2} interacts with the enzyme, δ–aminolevulinate (ALA) dehydrase to inhibit the formation of porphobilinogen. Pb^{+2} probably competes with the native Zn^{+2} centre required for the activity. The level of the δ–ALA gives a measure of Pb content in blood. Pb content in blood higher than 0.8 ppm will induce anemic conditions.

Pb^{+2} can interfere with Ca^{+2} and consequently bones are affected in poison. It can damage the mitochondria of kidney, allow the loss of glucose amino acids and phosphate through urine. Pb^{+2} can damage the liver and gastrointestinal treat. The toxicity of lead are–

muscle pain, join pain, depression, multiple sclerosis, encephalitis, behavioural problems, reproduction problems and tetragenic effect.

Conclusion

The drinking water of Lumding Town is not affected by Cadmium and Lead metal as observed by Atomic Absorption Spectrophotometer. Of course care should be taken from time to time to study the heavy metals because pollution of heavy metal (HM) *i.e.*, by Cadmium and Lead detriments the human health and increases the various types of communicable diseases. An integrated management work is taken to study and to reduce the various parameters present in excess water for drinking and for other purposes.

References

APHA (1995). *Standard Methods for Examination of Water and Wastewater*, 19th Edition. American Public Health Association, Washington DC, USA.

Das, A.K. (2004). *Inorganic Chemistry–Biological and Environmental Aspects*. Books and Allied (P) Ltd., Kolkata.

Dixit, R.C., Verma, S.R., Nitnaware, V. and N.P. Thaker (2003). Heavy metals contamination in surface and Ground water. *IJEH*, 45(2): 107–112.

Garg, A., Kripalani, C. and U. Brighu (2004). Heavy metals in the waters of a wastewater Drain (Amanishah Nallha) Near Mansarover Region in Jaipur and Ground waters in the vicinity. *Nat. Env. Poll. Techn.*, 3(3): 73–75.

Hoo, L.S., Samat, A. and M.R. Othman (2004). The level of selected Heavy metal (Cd, Cu, Fe, Pb, Mn and Zn) at residential Area Nearby Labu River System Riverbank, Malaysia. *Res. J. Chem. Env.*, 8(3): 24–29.

Maity, P.B., Sinha, T. and P.B. Ghosh (2003). Level of occurrence of some Metals in water of the River Hooghly within Kolkata Metropolitan Area. *Res. J. Chem. Env.*, 7(4): 17–19.

Chapter 20

Impact of Cadmium Chloride on the Behaviour and AChE Activity of a Freshwater Fish, *Oreochromis mossambicus*, Peters

Lingaraj Patro, Meenakshi Mishra and A.K. Panigrahi

ABSTRACT

The present study was designed to study effects of Cadmium chloride on a fresh water fish, Oreochromis mossambicus, Peters and its ecological implications. Exposed fishes appeared lethargic when compared to the control fish. Inappetence and ataxia was observed in the exposed fish. Exposed fishes showed erratic movements, loss of equilibrium, gradual onset of inactivity etc. when compared to the control fish. Significant changes were noted in AChE activity in brain, liver and muscle of the cadmium chloride exposed fish, when compared to control fish. The brain and liver was affected the most. Significant variations were marked between the control fish tissues and exposed fish tissues. These variations were due to the toxicant, Cadmium chloride. Partial insignificant recovery was noted during recovery studies. This enzyme is responsible for synaptic transmission and nerve impulse generation. This toxicant caused disturbance in nerve impulse generation and synaptic transmission in tissues.

Keywords: *Fish, Behaviour, AChE activity, Cadmium chloride.*

Introduction

Cadmium was recognised many years ago to be a highly toxic element but it was not until comparatively recently that concern began to be expressed over the possible effects on human health on long term exposure to low concentrations of this element. Cadmium is toxic to most living organisms. Interest in cadmium contamination began after the outbreak of itai-itai diseases in Japan. After exposure, the kidney is the organ, which contains the highest concentrations of cadmium and retains it longest. Moreover studies carried out in humans not occupationally exposed to cadmium reveal that 50 per cent body burden is found in the Kidneys (Kjellstrom, 1979). In recent years, much research has focused on determining factors contributing to the level of exposure and degree of accumulation of cadmium in renal cortex (Spicket and Lazner, 1979; Blanusa *et al.*, 1985 and Scott *et al.*, 1987). Cadmium in the renal cortex increases with age, reaching a maximum between 40–50 years (Elinder, 1985).

Cadmium is one of the most toxic metals occurring in nature. There is ample evidence in the literature that this metal is highly toxic to all living organisms (Yannai and Berdicevsky, 1995). In humans, chronic exposure to cadmium causes, severe damage to the kidneys and it has been linked to enhanced ageing processes, as well as cancer (Jentsch *et al.*, 1993). The toxic effects of cadmium are presumably associated with the metal's affinity for organic ligands containing sulphur, nitrogen or other electronegative functional groups. The most serious effects of Cadmium poisoning usually involve damage to the kidney, particularly the renal tubes. These symptoms are associated with proteinurea, glucosuria and high alkaline phosphatase in the blood. Chronic toxicity of cadmium is evident at 30-60 ppb in fresh water fish, 5 to 50 ppb at marine fishes. At high levels, cadmium causes kidney problems, anemia and bone marrow disorders. The major portion of cadmium ingested into our body is trapped in the kidneys and eliminated. A small fraction is bound most effectively by the body proteins, metallothionein, present in the kidneys, while the rest is stored in the body and gradually accumulates with age. When excessive amounts of Cd^{2+} are ingested, it replaces Zn^{2+} at key enzymatic sites, causing metabolic disorders. Keeping in view; the availability of heavy metal cadmium in the environment and their possible effect on the fresh water fishes surviving in the water bodies, this project was designed to evaluate the impact of cadmium on the toxicity, and acetylcholinesterase activity of a freshwater fish, *Oreochromis mossambicus*, Peters.

Materials and Methods

Oreochromis mossambicus, Peters were collected from the local Nursery, Berhampur, acclimatised and maintained in the laboratory, as described in Panigrahi and Misra (1980). Experiments were conducted at sub-lethal dose deduced from toxicity study. The MAC value of cadmium chloride was found to be 1.05 $mg.l^{-1}$ for 30 days and to be on the safe side 1.00 $mg.l^{-1}$ was considered for 28 days of the exposure and 28 days of recovery for sub-lethal studies. Both the control and exposed fishes were sacrificed at a 7 day interval. Brain, liver and muscle tissues were dissected out carefully, weighed and analysed for acetylcholinesterase activity using Rappaport *et al.* (1959) as detailed in Sigma technical Bulletin No. 420 (1977) technique. The AChE activity was expressed as μ/moles of ACh hydrolysed mg^{-1} of tissue hr^{-1}.

Results

All the exposed fish appeared lethargic after exposure to Cadmium chloride. The major clinical symptoms such as inappetance and ataxia appeared after 2 to 3 days exposure. At higher concentration of the Cadmium chloride, the exposed fish showed erratic movements. The other signs of toxicity such as loss of equilibrium, gradual onset of inactivity, erratic swimming with irregular collision to the inner glass walls of the aquarium was observed. Restlessness and least feeding were observed. The exposed fish shed fins and disintegration of dermal mucous layer was marked. Ulceration of the gill lamellae was observed in the exposed fish. Some symptoms in the exposed fish were similar to the preliminary symptoms of Epizootic Ulcerative Syndrome (EUS). Infection of eyes and bleeding of gills were not marked. Exophthalmia and involusions of test fish were not observed. Exposed fish could not regain their pre-exposed activity after transferring the exposed fish to toxicant free water in recovery studies. In the present investigation, peculiar symptoms were marked. The cadmium chloride exposed fish showed initial symptoms like mercury poisoning but later on the exposed fish became over excited and periodic outbursts in erratic swimming were noted, ultimately the excited fish died of suffocation. When the excited fishes were heavily oxygenated and kept in undisturbed water survived. The movement of the fish slowed down and became sluggish. The exposed fishes showed high level of sensitivity. Under the influence of any stimulus, the sluggish fish instantly started vigorous movement and many a times collided with the inner glass of the aquarium frequently and showed circular movements. After 1/2 minute, the movements slowed down and the fish became senseless, ventilation rate almost stops. If heavily oxygenated, the fish survives otherwise the exposed fish dies. During this later period of slowing down the movement, the colour of the fish becomes black due to dispersion of melanin pigment. This black colour disappears, when the exposed system was heavily oxygenated. At times, the shocked fish automatically returns to normalcy, and the black colour disappears. This type of self-recovery was only in 20 per cent of the cases. Heavy oxygenation can help to around 60 per cent suffered fishes to survive but the rest of the excited fish die instantly and never recover from the shock. In cadmium exposed fish lesions developed on the fins and later the skin erodes and the bones were clearly visible and shedding of fins were noted in few exposed fish towards later period of exposure or at higher exposure period. This seems to be an important symptom in cadmium exposure, which was not marked in mercury or pesticide exposure.

The changes in AChE activity in brain, liver and muscle of control and cadmium chloride exposed fish at different days of exposure and recovery and it's percent changes were shown in Table 20.1. The brain tissue showed a maximum decrease by 58 per cent and the enzyme activity declined from 0.386 ± 0.052 to 0.162 ± 0.048 µmole of ACh hydrolysed hr^{-1} mg^{-1} of tissue on 28th day of exposure (Table 20.1). When the exposed fish was transferred to toxicant free medium, the AChE activity declined from 0.385 ± 0.045 to 0.146 ± 0.035 µmole of ACh hydrolysed hr^{-1} mg^{-1} of tissue on 14th day of recovery and from 0.382 ± 0.042 to 0.152 ± 0.025 µmole of ACh hydrolysed hr^{-1} mg^{-1} of tissue on 28th day of recovery. This enzyme activity decreased by 62 per cent on 14th day recovery and decreased by 60.2 per cent over the 28th day exposure value, after 28 days of recovery, indicating no significant recovery at all. The percent decrease increased with the increase in exposure period from 7th day to 28th day. Initially the percent decrease was 16 per cent on 7th day, 31.4 per cent on 14th day, 54.1 per cent on 21st day and the highest of 58 per cent being on 28th day of exposure. The percent decrease increased significantly with the increase in exposure period in exposed fish exposed to cadmium chloride. The exposed fish brain enzyme activity could not recover even partly on 14th and 28th day of recovery showing permanent damage caused to the system. The changes in total AChE activity in liver of control and cadmium chloride exposed fish at different days of exposure and recovery and it's percent changes were shown in Table 20.2. The liver tissue showed a maximum decrease by 38.8 per cent and the enzyme activity declined from 0.268 ± 0.035 to 0.164 ± 0.044 μmole of ACh hydrolysed hr^{-1} mg^{-1} of tissue on 28th day of exposure. When the exposed fish was transferred to toxicant free medium, the AChE activity declined from 0.270 ± 0.045 to 0.168 ± 0.055 μmole of ACh hydrolysed hr^{-1} mg^{-1} of tissue on 14th day of recovery and from 0.269 ± 0.055 to 0.172 ± 0.046 μmole of ACh hydrolysed hr^{-1} mg^{-1} of tissue on 28th day of recovery. This enzyme activity decreased by 37.7 per cent on 14th day recovery and decreased by 36 per cent over the 28th day exposure value, after 28th days of recovery, indicating partial insignificant recovery. The percent decrease increased with the increase in exposure period from 7th day to 28th day. Initially the percent decrease was 3.4 per cent on 7thday, 17 per cent on 14th day, 31.2 per cent on 21st day and the highest of 38.8 per cent being on 28th day of exposure. The percent decrease increased significantly with the increase in exposure period in exposed fish exposed to cadmium chloride. The exposed fish liver enzyme activity could recover insignificantly by 1.1 per cent on 14th and 2.8 per cent recovery on 28th day of recovery showing drastic damage caused to the system. The changes in AChE activity in muscle of control and cadmium chloride exposed fish at different days of exposure and recovery and it's percent changes were shown in Table 20.2. The muscle tissue showed a maximum decrease by 32.7 per cent and the enzyme activity declined from 0.273 ± 0.023 to 0.162 ± 0.034 μmole of ACh hydrolysed hr^{-1} mg^{-1} of tissue on 28th day of exposure. When the exposed fish was transferred to toxicant free medium, the AChE activity declined from 0.274 ± 0.048 to 0.165 ± 0.042 μmole ofACh hydrolysed hr^{-1} mg^{-1} of tissue on 14th day of recovery and from 0.275 ± 0.036 to 0.185 ± 0.024 μmole of ACh hydrolysed hr^{-1} mg^{-1} of tissue on 28th day of recovery. This enzyme activity decreased by 39.7 per cent on 14th day recovery and decreased by 32.7 per cent over the 28th day exposure value, after 28 days of recovery, indicating partial recovery. The percent decrease increased with the increase in exposure period from 7th day to 28th day. Initially the per cent decrease was 10.9 per cent on 7th day, 25.1 per cent on 14th day, 32.7 per cent on 21st day and the highest of 40.6 per cent being on 28th day of exposure. The percent decrease increased significantly with the increase in exposure period in exposed fish exposed to cadmium chloride. The exposed fish muscle enzyme activity could recover only partly by 1.3 per cent on 14th and by 7.9 per cent on 28th day of recovery showing damage caused to the system. Out of the three tissues studied, the exposed brain showed the highest damage than liver and muscle. The correlation coefficient analysis between days of exposure and AChE activity in control fish brain showed a positive but non–significant correlation, whereas, the exposed fish brain showed a significant negative correlation ($r = -0.989$, $P \leq 0.01$). The control set did not show any positive significant correlation (P = NS). The percent change in the enzyme activity in the exposed fish brain when compared to control fish brain showed the existence of a negative significant correlation with the exposure period. The correlation coefficient analysis between days of exposure and the AChE activity of liver slices of the control fish did not show any significant correlation, whereas the exposed fish liver showed a negative and significant correlation ($r = -0.987$, $P \leq 0.01$) with the exposure period. The percent change in AChE activity showed the existence of a negative, significant correlation ($r = -0.984$, $P \leq 0.01$) with the exposure period. The analysis of variance ratio test indicated the existence of significant difference between rows and non-significant difference between columns. The control muscle did not show any significant correlation between exposure period and enzyme activity. However, the muscle of the exposed fish showed negative and significant correlation ($r = -0.985$, $P \leq 0.05$) with the exposure period. The percent change in the enzyme activity showed negative and significant correlation with the exposure period. The analysis of variance ratio test for muscle showed significant difference between rows and non-significant difference between columns. The exposed fishes were when transferred to toxicant free medium; the enzyme activity showed recovery in liver and muscle tissues but the recovery was not statistically significant. Brain showed the highest damage and no recovery, liver showed the lowest recovery and muscle showed partial recovery after 28 days of recovery.

Table 20.1: Changes in Total AChE Activity (µmoles of ACh hydrolysed mg^{-1} hr^{-1}) in Control and Cadmium Chloride Exposed Fish Brain, Liver and Muscle at Different Days of Exposure and Recovery

Tissue	*Condition*	*Exposure in Days*					*Recovery in Days*	
		0	*7*	*14*	*21*	*28*	*14*	*28*
Brain	Control	0.385±0.065	0.375±0.041	0.385±0.062	0.382±0.065	0.386±0.052	0.385±0.045	0.382±0.042
	Exposed	0.385±0.065	0.315±0.045	0.264±0.042	0.175±0.035	0.162±0.048	0.146±0.035	0.152±0.025
Liver	Control	0.268±0.070	0.264±0.045	0.270±0.026	0.269±0.065	0.268±0.035	0.270±0.045	0.269±0.055
	Exposed	0.268±0.070	0.255±0.042	0.224±0.022	0.185±0.054	0.164±0.044	0.168±0.055	0.172±0.046
Muscle	Control	0.276±0.026	0.275±0.035	0.274±0.055	0.275±0.012	0.273±0.023	0.274±0.048	0.275±0.036
	Exposed	0.276±0.026	0.245±0.056	0.205±0.026	0.185±0.028	0.162±0.034	0.165±0.42	0.185±0.024

Table 20.2: Percent Change in Total AChE Activity in Exposed Fish Brain, Liver and Muscle at Different Days of Exposure and Recovery, when Compared to Control Tissues

Tissue	*Exposure in Days*					*Recovery in Days*	
	0	*7*	*14*	*21*	*28*	*14*	*28*
Brain	0.0	–16.0	–31.4	–54.1	–58.0	–62.0 (NR)	– 60.2 (NR)
Liver	0.0	–3.4	–17.0	–31.2	–38.8	–37.7 (1.1)	–36.0 (2.8)
Muscle	0.0	–10.9	–25.1	–32.7	–40.6	–39.7 (1.3)	–32.7 (7.9)

Values in parentheses indicate percent recovery.

Discussion

The toxicity of cadmium chloride becomes more apparent in a very shorter period in aquatic animals. The observed depression in active metabolism in cadmium exposed fish were indicative of damage to nervous tissues, inhibition of enzymes or a vital system, was totally in agreement with Panigrahi (1980). The route of entry of the toxicant is generally agreed to be via the gills and thus gets directly into the vascular system. This causes damage to tissues and as a result of which there is depression in active metabolism (Mac Leod and Pessah, 1973). Depression in active metabolism that directly reduced scope of activity of the animal was demonstrated by Fry (1957). Probably this has relevance with the impaired swimming ability, erratic movements etc. which were observed in all the test fish. High percentage mortality of fish due to action of the cadmium chloride might be due to the pathological changes. Because of its high toxicity, most countries include cadmium among the "Priority pollutants" requiring suitable treatment prior discharge into the environment (Puranik *et al.*, 1995). The United States Environmental Protection Agency limits Cadmium levels in drinking water to 0.001 mg/l. In India, the permissible concentration of Cadmium in the industrial effluents discharged into inland surface waters is 0.1 mg/l^{-1}. Harichandan (2002) reported that loss of appetite, loss of weight, nervousness, dizziness, loss of equilibrium, erratic swimming and gradual onset of inactivity were the sub-clinical effects of cadmium intoxication. Identical observations were not marked here in *Oreochromis* (*Tilapia*) fish exposed to cadmium chloride. The cadmium chloride exposed fish showed initial symptoms like mercury poisoning but later on the exposed fish became over excited and periodic outbursts in erratic swimming were noted, ultimately the excited fish died of suffocation. When the excited fishes were heavily oxygenated and kept in undisturbed water survived. The movement of the exposed fish slowed down and became sluggish. The exposed fishes showed high level of sensitivity. Under the influence of any stimulus, the sluggish fish instantly starts vigorous movement and many a times colloid with the glass of the aquarium frequently and showed circular movements. After 1/2 minutes, the movements slow down and the fish becomes senseless, ventilation rate almost stops. If heavily oxygenated, the fish survives otherwise the exposed fish dies. During this later period of slowing down the movement, the colour of the fish becomes black due to dispersion of melanin pigment. This black colour disappears, when the system was heavily oxygenated. At times, the shocked fish automatically comes back to normal, and the black colour disappears. This type of self recovery was only in 30 per cent of the cases. Heavy oxygenation can help to around 65 per cent fishes to survive but the rest of the excited fish die instantly and never recovered from the shock. In cadmium exposed fish lesions developed on the fins and later the skin erodes and the bones were clearly visible and becomes naked and shedding of fins were noted in few exposed fish towards later period of exposure or at higher exposure period (Harichandan, 2002). This seems to be an important symptom in cadmium exposure, which was not marked in mercury or pesticide exposure.

In the preliminary study the animals showed all regular features of cadmium poisoning; such as excitation, irritation and restlessness. Towards the end of the experimental period *i.e.* prior to death the locomotion of fish almost ceased and remain suspended vertically in the water medium, indicated loss of equilibrium. In addition, periodic and erratic paralytic movements were observed in exposed fish. The early signs of poisoning were probably due to the effect of chemical cadmium on fish. The body of the exposed fishes did not show semi U or V bending either in the earlier period or towards later period of exposure, like mercury poisoning as reported by Panigrahi and Misra (1979, 80). Control fish remained clinically healthy, throughout the experimental period. The observed depression in active metabolism in cadmium chloride exposed fish are indicative of damage to nervous tissues, inhibition of enzymes or vital system was totally in agreement with Mac Leod and Pessah (1973); Panigrahi (1980) and Misra (2002). Fry (1957) demonstrated that depression in active metabolism directly reduce "scope for activity". This in turn may result in decreased growth, impaired swimming ability with erratic outbrust at times. It is possible that even a small reduction in the normal active metabolic rate may be indicative of stress in the animals (Mac Leod and Pessah, 1973). Out of the three tissues studied, the exposed brain showed the highest damage than liver and muscle. The correlation coefficient analysis between days of exposure and AChE activity in control fish brain showed a positive but non-significant correlation, whereas, the exposed fish brain showed a significant negative correlation ($r = -0.974$, $P \leq 0.01$). The correlation coefficient analysis between days of exposure and the AChE activity of liver slices of the control fish did not show any significant correlation, whereas the exposed fish liver showed a negative and significant correlation with the exposure period. The control muscle did not show any significant correlation between exposure period and enzyme activity. However, the muscle of the exposed fish showed negative and significant correlation with the exposure period. From the results obtained in this study and the reports of Harichandan (2002), it is clearly evident that, cadmium chloride, which is now well known for its interference in biochemical metabolism, becomes toxic when used in lower and moderate concentrations. Toxicant build up in the plants and consequent build up in grazers and in fish may lead to availability of toxicant in milk and meat of herbivores or grazers. From milk, fish, meat and rice grain, the toxicant may enter into human body. This build up of toxicant in the human system may lead to another significant incident, like many other incidents related to cadmium pollution. The route of entry of cadmium to brain is mostly by blood vascular system. The excretion rate of cadmium in brain is very slow, as it lacks any special mechanism of excretion (Samant, 1989). Larman *et al.* (1976) reported that by removing the fish from contaminated areas and placing them in uncontaminated environments do not reduce the cadmium concentration but rather get diluted due to the growth of the fish. The extreme behaviour of the exposed fish, exposed to cadmium containing waste, now can be correlated with the enzyme activity. There exists a strong correlation among the ions concentration in the tissue, ATPase activity, AChE activity and metabolic rate (Panigrahi, 1980; Webb, 1966 and Harichandan, 2002) with the behaviour of the fish. The report of Harichandan (2002) strongly indicates that the enzyme activities are related to residual cadmium concentration in the tissues. We can conclude here that probably the causative agent for all such observed changes in this study is the cadmium present in the aquarium.

Acknowledgement

The authors wish to thank the Head, Department of Botany for laboratory facilities. The financial assistance given by UGC (Eastern Region) Kolkata is thankfully acknowledged.

References

Blanusa, M., Kralj, Z. and A. Bunarevic (1985). Interaction of Cadmium, Zinc and Copper in relation to smoking habit, age and histopathological findings in humans kidney cortex. *Arch. Toxicol.*, 58: 115–117.

Elinder, C.G., Edling, C., Lindberg, E., Kagedas, B. and O. Vesterberg (1985). B_2-Microglobulinuria among workers previously exposed to Cadmium: Follow-up and dose response analyses. *Am. J. Ind. Med.*, 8: 553–564.

Fry, F.E.J. (1957). The aquatic respiration of fish. In: *The Physiology of Fish*, (Ed.) M.E. Brown. Academic Press, New York, 1: 1.

Harichandan, N.L. (2002). Toxicological effects of cadmium chloride contained effluent on a fresh water fish and its ecological significance. *Ph.D. Thesis*. Berhampur University, Orissa.

Jentsch, S., Schobert, C., Reins, H.A. and Jungman (1993). Resistance to cadmium mediated by ubiquitin-dependent proteolysis. *Nature*, 361: 369–371.

Kjellstrom, T. (1979). Exposure and accumulation of cadmium in populations from Japan. *The United States and Sweden. Environ. Health Perspect.*, 28: 169–197.

Larman, P., W. Wayne, A. Willford and James R. Olson (1976). Transections of American Fisheries Society, 105 (2): 296.

MacLeod, J.C. and E. Passah (1973). Temperature effects on mercury accumulation, toxicity and metabolic rates in rainbow trout (*Salmo gairdeneri*). *J. Fish. Res. Bd. Canada*, 130: 485.

Mishra, M. (2002). Physiological changes induced by red mud leached waste of NALCO, Damonjodi on a fresh water fish and its ecological implications. *Ph.D. Thesis*. Berhampur University, Orissa.

Panigrahi, A.K. (1980). Eco-physiological effects of mercurial compounds on some fresh water fish (Toxicological effects of inorganic mercury [$Hg(NO_3)_2$] on two fresh water fishes, *Tilapia mossambica*, Peters and *Anabas scandens*. Cuv. and Val. and its ecological implications. *Ph.D. Thesis*. Berhampur University.

Panigrahi, A.K. and B.N. Mishra (1978). Toxicological effects of inorganic mercury on a fresh water fish, *Anabas scandens*, Cuv, and Val. and their ecological implications. *Env. Pollut.*, 16(1): 31.

Panigrahi, A.K. and B.N. Mishra (1980). Toxicological effects of inorganic mercury on a fresh water fish, *Tilapia mossambica*,

Peters and their ecological implications. *Arch. Toxicol.*, 44: 269–278.

Puranik, P.R., Chabukswar, N.S. and Paknikar, K.M. (1995). Cadmium biosorption by *Streptomyces pimprina* waste biomass. *Appl. Microbiol. Biotechnol.*, 43: 1118–1121.

Rappaport, F., Fischl, J. and N. Pinto (1959). An improved method for the estimation of cholinesterase activity in serum. *Clin. Chim. Acta*, 4: 227.

Samant, D.P. (1987). Toxicological effects of a mercury based fungicide on a fresh water fish and its ecological implications. *Ph.D. Thesis.* Berhampur University.

Scott, R., Aughey, E., Fell, G.S. and M.J. Quinn (1987). Cadmium concentrations in human kidneys from the UK. *Hum. Toxicolk.*, 6: 111–120.

Sigma Technical Bulletin (1977). The colorimetric determination of cholinesterase in serum or plasma. 420: 1–9.

Spicket, J.T. and J. Lazner (1979). Cadmium concentration in human kidney and liver tissues from Western Australia. *Bull. Environ. Contam. Toxicol.*, 23: 627–630.

Webb, J.L. (1966). Enzymes and metabolic inhibitors, 1st ed., Vol. II. Academic Press, New York, pp.729.

Yannai, S. and I. Berdicevsky (1995). Formation of organic cadmium by marine microorganisms. *Eco-toxicology and Environmental Safety*, 32: 209–214.

Chapter 21

Bioaccumulation of Trace Metals in Marine Algae (Chaetomorpha) at Tharangambadi Coast, South East Coast of India

P. Martin Deva Prasath

ABSTRACT

Growing industrialization and population explosion are the major factors responsible for marine pollution. Disposal of industrial, anthropogenic and municipal wastes into the neighbouring streams and rivers causes many health hazards. This plays a vital role in raising the levels of heavy metals in the marine environment. Hence, monitoring seasonal variations in the concentrations of heavy metals has become an important concern of the environmental researchers. Therefore, this study is aimed to assess the trace metal concentration (Zn, Cu, Fe and Mn) in water, sediments and marine algae (Chaetomorpha) of Tranquebar coast in Nagapattinam District of Tamil Nadu.

Keywords: *Trace metals, Marine algae, Tharangambadi, Bioaccumulation, South East Coast, India.*

Introduction

Growing industrialization, population explosion and migration to urban centers are major factors responsible for environmental pollution. The disposal of anthropogenic wastes, dumping of municipal and industrial wastes in the neighbouring streams and rivers have quite significant role in raising the level of toxic metals in coastal areas. Metals are biologically non-degradable and through food-chain, they may finally pass on to man (Thomas and Jaquet, 1976). Numerous heavy metal studies have been made to understand the effect of heavy metals on animals, but relatively little attention has been paid to study the trace metal accumulation on seaweeds. Hence, along with the marine algae, Chaetomorpha, accumulation of trace metals (Zn, Cu, Fe and Mn) in water and sediment in Tharangambadi coast were also studied.

Materials and Methods

Samples of water, sediments and *Chaetomorpha* were collected at monthly interval in Tharangambadi coast (Lat. 11°2'N; Long. 79°49'E) in Nagapattinam district of Tamil Nadu, South India for a period of two years from January 2000 to December 2001. Tharangambadi coast receives domestic and municipal sewages, shrimpfarm effluents and agricultural discharges mainly through Uppanar estuary of Tharangambadi. The following methods are adopted for the detection of Zn, Cu, Fe and Mn in Water, Sediments and Marine Algae.

Water (Brooks *et al.*, 1967)

Surface water samples were collected in precleaned and acid washed polypropylene bottles and the samples were filtered in Millipore filter paper (Pore size 0.45µ). The samples were preconcentrated with APDC-MIBK extraction procedure (Brooks *et al.*, 1967). The resulting solution was aspirated to the Flame Atomic Absorption Spectrophotometer (Perkin-Elmer model 373) for the detection of Zn, Cu, Fe and Mn in water.

Sediment and Algae (Chester and Hughes, 1967)

Sediments samples were collected in precleaned, acid washed PVC corer and the algae Chaetomorpha were collected by hand picking in intertidal Tharangambadi coast for a period of two years from January 2000 to December 2001. Both the samples were washed with metal free double distilled water and dried in hot air oven at 110°C for 5–6 hrs and ground to powder in a glass mortar and stored in precleaned polythene bags. 500mg of sample was taken and digested with a mixture of 1 ml of Con H_2SO_4, 5 ml of Con HNO_3 and 2 ml of $HClO_4$. A few drops of hydrofluoric acid is added to achieve complete digestion and filtered the sample to make up to 25 ml with metal free double distilled water for the estimation of Zn, Cu, Fe and Mn using Flame Atomic Absorption Spectrophotometer (Perkin-Elmer model 373).

Results

Monthly variation in the bioaccumulation of Zn, Cu, Fe and Mn in water, sediments and marine algae recorded at Tharangambadi Coast are shown in Figures 21.1–21.12. The annual mean concentration of the metals was also calculated.

Monthly variation in the accumulation of Zn in water varied from 4.23 μgl^{-1} (May 2000) to 102 μgl^{-1} (September 2000); Cu varied from BDL (Below Detectable Level, July and August 2001) to 40.2 μgl^{-1} (April 2000); Fe varied from 84.82 μgl^{-1} (May 2000) to 260.4 μgl^{-1} (February 2000); and Mn varied from 0.82 μgl^{-1} (January 2000) to 47.6 μgl^{-1} (May 2000) during the year January 2000 to December 2001.

Monthly variation in the accumulation of Zn in sediment varied from 10 μgg^{-1} (May 2000) to 192.2 μgg^{-1} (December 2001); Cu varied from BDL (February 2000) to 174 μgg^{-1} (December 2000); Fe varied from 343 μgg^{-1} (January 2001) to 20800 μgg^{-1} (February 2000); and Mn varied from 79 μgg^{-1} (December 2000) to 881 μgg^{-1} (February 2001) during the year January 2000 to December 2001.

Monthly variation in the accumulation of Zn in Chaetomorpha varied from 7.5 μgg^{-1} (August 2000) to 68.1 μgg^{-1} (October 2001); Cu varied from 9.5 μgg^{-1} (July 2000) to 69 μgg^{-1} (September 2001); Fe varied from 381.5 μgg^{-1} (July 2000) to 1274.5 μgg^{-1} (September 2001); and Mn varied from 43 μgg^{-1} (May 2000) to 392 μgg^{-1} (November 2001) during the year January 2000 to December 2001.

Discussion

The concentration of Zn, Cu, Fe and Mn in water was high during monsoon season and low during summer season. The high values of Zn, Cu, Fe and Mn during monsoon season are due to disposal of agricultural, domestic, mur.icipal and shrimp farm discharges. Further, natural sources of heavy metals are through land and river runoff and from mechanical and chemical weathering

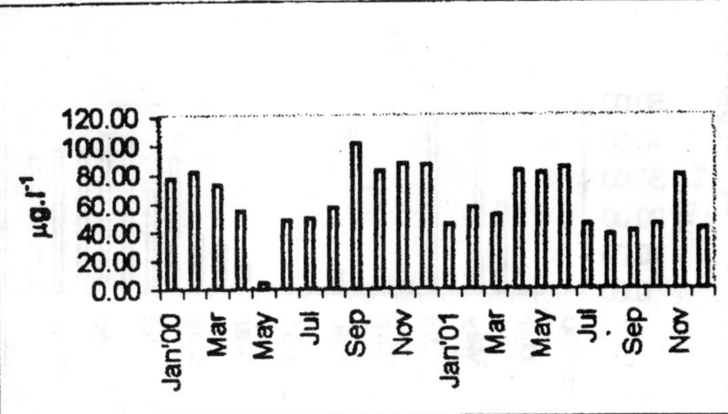

Figure 21.1: Monthly Variations in Zn Concentration (µg.l^{-1}) in Water Recorded at Tharangambadi Coast During January 2000 to December 2001

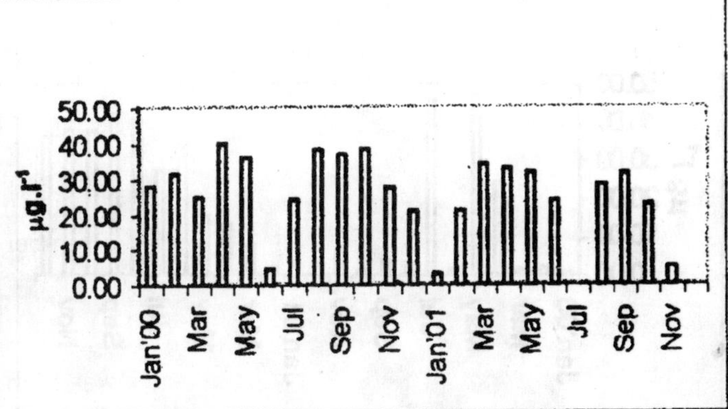

Figure 21.2: Monthly Variations in Cu Concentration (µg.l^{-1}) in Water Recorded at Tharangambadi Coast During January 2000 to December 2001

WATER

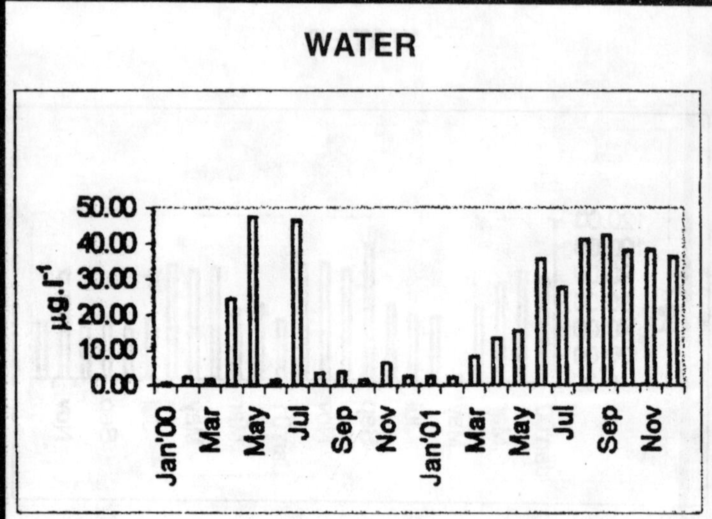

Figure 21.3: Monthly Variations in Fe Concentration (µg.l^{-1}) in Water Recorded at Tharangambadi Coast During January 2000 to December 2001

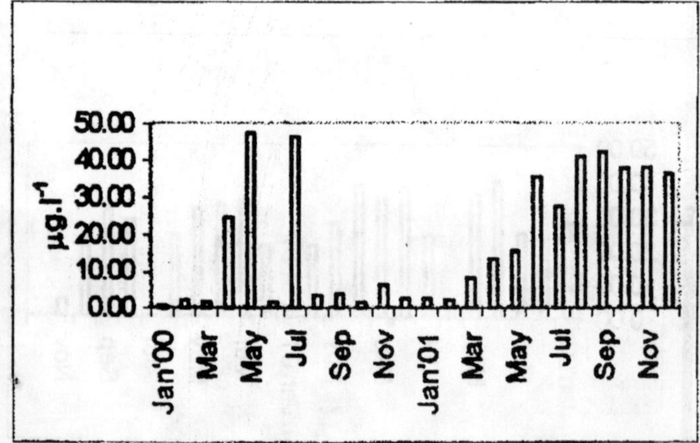

Figure 21.4: Monthly Variations in Mn Concentration (µg.l^{-1}) in Water Recorded at Tharangambadi Coast During January 2000 to December 2001

SEDIMENTS

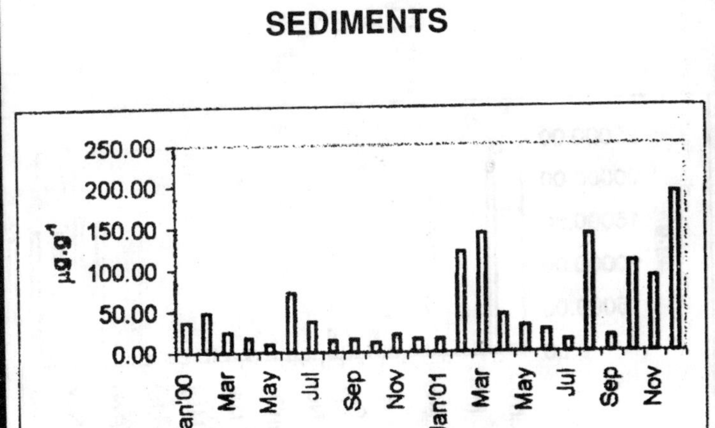

Figure 21.5: Monthly Variations in Zn Concentration (µg.g^{-1}) in Sediments Recorded at Tharangambadi Coast During January 2000 to December 2001

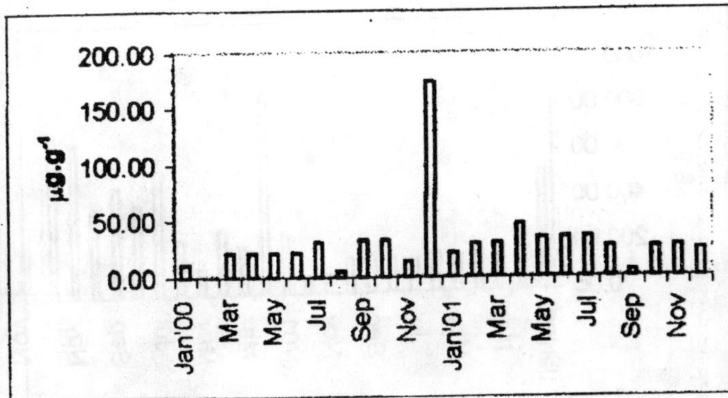

Figure 21.6: Monthly Variations in Cu Concentration (µg.g^{-1}) in Sediments Recorded at Tharangambadi Coast During January 2000 to December 2001

SEDIMENTS

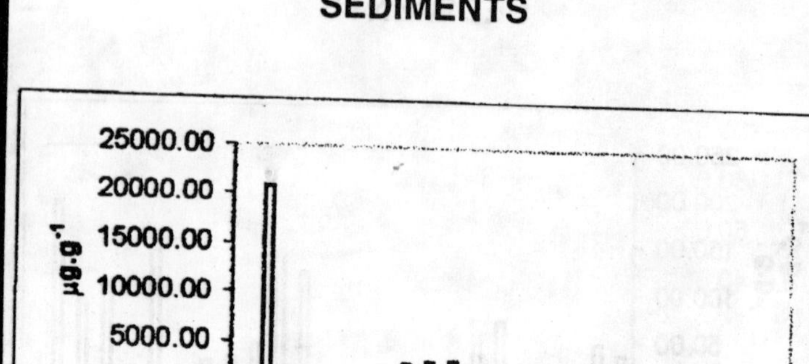

Figure 21.7: Monthly Variations in Fe Concentration ($\mu g.g^{-1}$) in Sediments Recorded at Tharangambadl Coast During January 2000 to December 2001

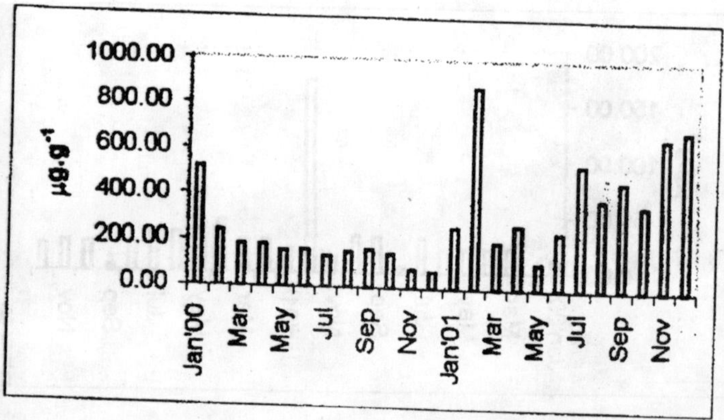

Figure 21.8: Monthly Variations in Mn Concentration ($\mu g.g^{-1}$) in Sediments Recorded at Tharangambadi Coast During January 2000 to December 2001

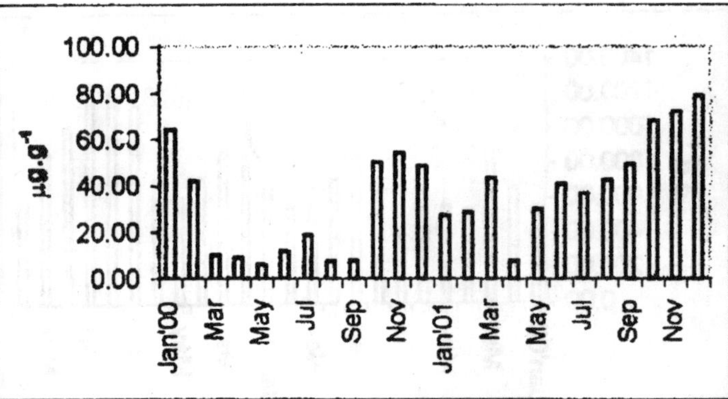

Figure 21.9: Monthly Variations in Zn Concentration ($\mu g.g^{-1}$) in Algae Recorded at Tharangambadi Coast During January 2000 to December 2001

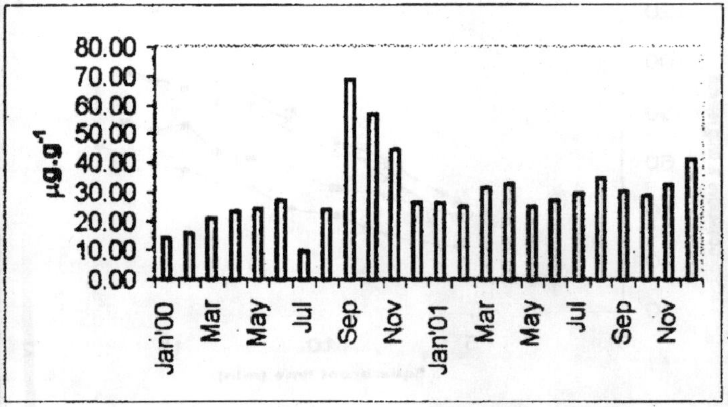

Figure 21.10: Monthly Variations in Cu Concentration ($\mu g.g^{-1}$) in Algae Recorded at Tharangambadi Coast During January 2000 to December 2001

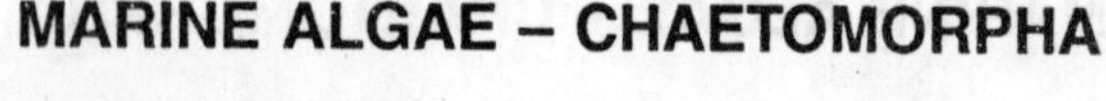

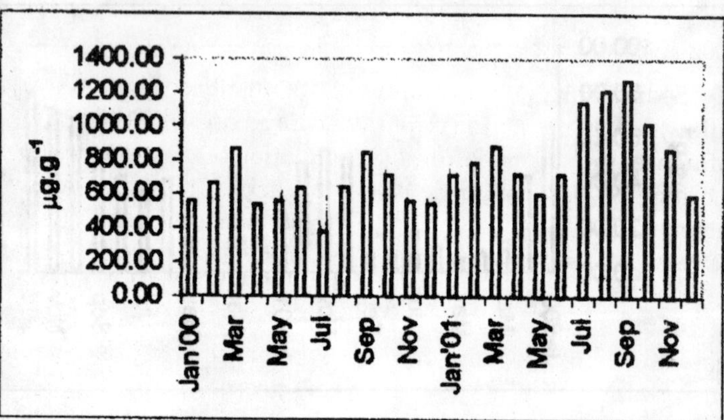

Figure 21.11: Monthly Variations in Fe Concentration ($\mu g.g^{-1}$) in Algae Recorded at Tharangambadi Coast During January 2000 to December 2001

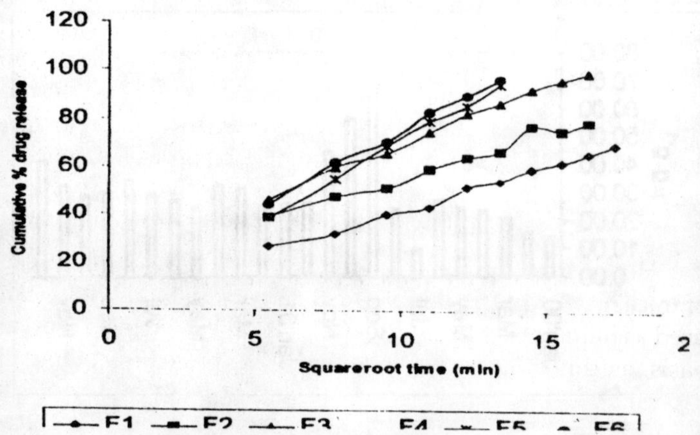

Figure 21.12: Monthly Variations in Mn Concentration ($\mu g.g^{-1}$) in Algae Recorded at Tharangambadi Coast During January 2000 to December 2001

of rocks [Bryan, 1984]. The low values of metals during summer and premonsoon season are due to utilization and absorption of metals by primary producers like algae and phytoplankton which showed high density due to stagnant environmental condition (Rajaram, 2002 and Sampathkumar *et al.*, 1992).

Sediment metal concentrations are mainly controlled by river inflow, sediment particle type and size and organic contrast which influences the absorption and accumulation of metals in sediments (Loring, 1978). The higher values of metals in sediment during monsoon seasons are due to land runoff and the influx of the metal rich fresh water increased the particulate matter and suspended sediment load along with shrimp farm effluents. The lower values of trace metals noted during summer season are due to low fresh water inflow, biological utilization, dominance of high saline water, precipitation of particulate matter, decreased land drainage, sediment particle size and type, mud content, sulphide content and microbial activity (Craig and Moreton, 1983; and Rajaram, 2002).

Marine algae are good accumulator of trace metals. Uptake and accumulation of trace metals by the algae depends on availability of elements in water, environmental condition and species composition and population size (Kumaraguru, 1980). In the present study, the high levels of trace metals Zn, Cu, Fe and Mn in algae during monsoon season are due to the pre existence of high concentration of metals found in water and sediment which are transferred to algae, low salinity and algal density in monsoon season increased the accumulation rate by adsorption and absorption mechanisms (Boyle *et al.*, 1976) and low values during premonsoon season may be due to lack of rainfall, resultant runoff and low algal density.

Hence, from the present study, based on the bio-accumulation of trace metals in Chaetomorpha, it is concluded that algae can be used as monitors of aquatic heavy metal pollution, on a long-term basis, as this species shows greater heavy metal accumulation capacity.

References

Boyle, E.A., Sclater, F.R. and J.H. Edmond (1976). On the marine geochemistry of cadmium. *Nature*, London, 263: 42–44.

Brooks, R.R., Presley, B.J. and I.R. Kalplan (1967). APDC-MIBK extraction system for the determination of trace metals in saline water by atomic absorption spectroscopy. *Talanta*, 14: 809–816.

Chester, R. and M.J. Hughes (1967). A chemical technique for the separation of ferromanganese materials, carbonate minerals and absorbed trace elements from pelagic sediments. *Chem. Geol.*, 2: 249–263.

Craig, P.J and P.A Moreton (1983). Total Hg, methyl Hg and sulphide in River Carron sediments. *Mar. Pollut. Bull.*, 14: 408–411.

Kumarguru, A.K. (1980). Studies on the chemical and biological transport of heavy metal pollutants Copper, Zinc and Mercury in Vellar estuary and the toxicity of these pollutants to some estuarine fish and shell fish. *PhD. Thesis*, Annamalai University.

Loring, D.H. (1978). Environmental biogeochemistry and geomicrobiology: Methods, metals and assessment. In: *Proceedings of 3rd International Symposium on Environmental Biogeochemistry* (Ed.) W.E. Krumbein. Wolfenbuttal, W. Germany, Ann. Arbor. Science Publication, Inc., Ann Arbor Michigam, p. 1025–1010.

Rajaram, R. (2002). Studies on hydrobiology and mercury content in Uppanar estuary. Cuddalore, Southeast coast of India. *Ph.D. Thesis*, A.U., India, pp. 128.

Sampathakumar, P. (1992). Investigation on plankton in relation to hydrobiology and heavy metals in the Tranquebar, Nagapattinam Coast. *Ph.D. Thesis*, A.U., India, pp. 184.

Thomas, R.L. and J.M. Jaquet (1976). *Fish Res. Board*, Canada, 33: 404.

Chapter 22

Acute Toxicity and Behavioural Responses to Nickel Sulphate to the Fish *Heteropneustes fossilis*

M. Choudhary and M.M. Jha

ABSTRACT

Release of heavy metals into atmosphere contaminate the aquatic system. Such contamination deteriorates the water quality and results into large scale fish mortality. Further more the toxic potential of a particular chemical reduces the fitness of a population. It was thus, felt necessary to have an accurate method for finding out toxic calculations of a given chemical to a given animal in a specific time. Acute toxicity test in form of LC_{50} *are the most widely practiced and accepted methods for the determination of such toxic levels. The toxicity of Nickel sulphate to H. fossilis was evaluated,by (Table 22.2) and (Table 22.3). The* LC_{50} *values obtained by former method were 425.3 mg/l, 382.9 mg/l, 351.1 mg/l and 333.6 mg/l for 24, 48, 72 and 96 hours where as the latter method gives* LC_{50} *values to be 450 mg/l, 400 mg/l, 380 mg/l and 334.8 mg/l for 24, 48, 72 and 96 hours respectively. Surfacing, rapid uppercular movements, convulsions, profuse mucus secretion's and its deposition over the gills gyratery, swimming movements by moribund fish and loss of equilibrium before death were the prominent behavioral changes noticed during present study (Table 22.1).*

Introduction

The nickel toxicity has assumed great interest because of the widespread environmental occurrence of nickel and the incidence of accidental parsoning in nickel workers (Sunderam *et al.*, 1988a). Though quite a good number of toxicity tests on nickel have been performed on various animals including fishes (Gill and Pant 1981; Saxena and Parshari, 1983; Khangarot and Ray, 1990; Jha and Jha 1995; Roy 2003). No study has yet been undertaken to determine the toxicity of nickel to *H. fossilis*, one of the most important for ever water fishes.

The present study has, therefore, been undertaken to establish the acute toxicity in term of LC_{50}, of Nickel sulphate to this fish in relation to the behavioural responses various methods like logit, regression equation, graphic etc. but these methods have been shown to have a number of statistical deficiencies (Roy, 1988), therefore an attempt has been made in the present study to evaluate the LC_{50} of Nickel sulphate by yet another but more accurate logistic response model of Byron and Brown (1967).The accuracy of the values so obtained have been confirmed and compared with the LC_{50} values, calculated by establishing regression lines following the method of Downie and Heath, 1970.

Materials and Methods

The adult living species of *H. fossilis* were collected from a local fish pond and were brought to the laboratory in wide–mouthed large earthen pots filled half with natural water. The fish were washed with 0.1 per cent known solution to remove dernal infections if any. Healthy fish of average length (9–11 cm) and weight (21–24 gm) were selected and acclimated in 40 litre glass aquaria to the laboratory conditions for a fortnight. Running tap water was used in all the experiments.

Bioassays were conducted for the determination of LC_{50} value of nickel sulphate ($NiSO_4$, $6H_2O$, BDH analytical reagent grade) for 24, 48, 72 and 96 hours following the methods of APHA, AWWA and WPCF (1985).

A stock solution of Nickel sulphate was prepared by similar dilution technique as described by Doudoroff *et al.* (1957). The fish were exposed to different concentrations of nickel sulphate and their percent survival at different time intervals (24, 48, 72 and 96 hrs), were recorded as a measure of acute toxicity.

Results

Table 22.1: Physico-chemical Characteristics of the Test Water

Characteristics	*Unit*	*No. of Observation*	*Mean*	*Range*
pH		10	7.7	7.2–7.9
Temperature	°C	10	23.6	24–28
Total solids	mg/l	10	37.5	30.6–38.5
Dissolved solids	mg/l	10	12.0	10.0–12.5
Dissolved oxygen	(DO) mg/l	10	6.0	5.0–7.3
Free CO_2	mg/l	5	2.0	1.2–3.0
Total hardness	mg/l as $CaCO_3$	5	175.0	160.0–200.0
Total alkalinity	mg/l as $CaCO_3$	5	150.0	136.0–165.0

Table 22.2: LC_{50} Values of Nickel Sulphate to *H. fossilis* Calculated by Logistic Response Model

Sl.No.	*Time Intervals (Hours)*	*Antilog*	*LC_{50} (mg/l)*
1.	24	242.6287	425
2.	48	2.5831	382.9
3.	72	2.5456	351.2
4.	96	2.5233	333.6

Table 22.3: LC_{50} Values of Nickel Sulphate to *H. fossilis* Calculated by Regression Equation

Sl.No.	*Time Intervals (Hours)*	*Regression Equation*	*LC_{50} Values*
1.	24	$-189 + 0.53x$	450 (mg/l)
2.	48	$-155 + 0.5x$	400 (mg/l)
3.	72	$-140 + 0.5x$	380 (mg/l)
4.	96	$-2930 + 8.9x$	334 (mg/l)

Discussion

The behavioural observations noticed during the present study are in agreement with those of Gill and Pant (1981), Saxena and Parashari (1983), Jha and Jha (1995), Roy (2003). The rapid opercular movements and culvelsions observed in the present case may be due to the hypotoxic condition created by nickel toxicity. Similar was also observed Lal *et al.* (1984). Secretion of mucus is a normal feature under stressful conditions. Jha and Jha (1995) also observed similar with Nickel chloride to *A. testudineus*. The gyratory movements as observed in the present study may be attributed to the pathogenic effect of nickel chloride such as fatigue and comatose behaviour (Roy,. 1988).The apparent loss of equilibrium in the nickel exposed fish is indicative of the fact that nickel also disturbs and influence the central nervous system. It is, therefore, calculated that the toxicity of nickel depends upon a number of physical, chemical and biological factors. Each of which may be used as a tool for the prediction of metal toxicity to fishes.

References

APHA, AWWA and WPCF (1985). *Standard Methods for the Examination of Water and Wastewater*, 16th edn. American Public Health Association, American Water Works Association and Water Pollution Control Federation, American Public Health Association Washington, D.C.

Byron, W. and J.R. Brown (1967). Quantal response assay. In: *Statistics in Endocrinology*, (Eds.) J.U. Mc Arthur and T. Colton. MIT Press, Cambridge, MA, London.

Doudoroff, P., Anderson, B.G. Burdick, G.E. Galtsoft, P.S., Hart, W.B., Partick,. R., Strong, E.K., Surber, E.W. and W.M. Vonhorn (1957). Bioassay methods for the evaluation of acute toxicity of industrial wastes to fresh sewage. *Ind. Wastes*, 23(11): 1380–1397.

Downie, N.M. and R.W. Heath (1970). *Basic Statistical Methods*, Harper International Edition, New York,.Evanston and London, p. 129–131.

Gill, T.S. and J.C. Pant (1981), Toxicity of nickel to the fish *Puntius Conchonius* and its effects on blood glucose and liver glycogen. *Com. Physiol. Ecol.*, 6(2): 99–102.

Jha, B.S. and M.M. Jha (1995). Acute toxicity of nickel chloride to freshwater climbing perch, Anabas testudineus (Bloch). *Coluban Journal of Life Science*, pp. 11, Figure 1, Table 2.

Khangarot, R and P.K. Ray (1990). Acute toxicity and toxic interaction of chromium and nickel to common gurpy *Poecilia reticulata* (Peters). *Bull. Environ. Contam. Toxicol.*, 44: 832–839.

Lal, H., Mishra, V. Vishwanathan, P.W. and C.R. Krishnamurty (1984). Effect of synthetic detergents on some of the behavioural, patterns of fish fingerlings (*Cirrhina Mrigala*) and its relation to ecotoxicology. *Bull. Environ. Contam. Toxicol.*, 32: 109–115.

Roy, D. (1988). Toxicity of an anionic detergent, dodecylbenzene sodium sulfonate, to a freshwater fish, *Rita rita*. Determination of LC_{50} values by different methods. *Ecotoxicol. Environ. Saf.*, 15: 186–194.

Roy, D.N. (2003). Studies on the effect of chromium to the fish, *H. fossilis*. *Ph.D. Thesis*, L.N. Mithila University, Darbhanga.

Saxena, O.P. and A. Parashari (1983). Comparative study of the toxicity of six heavy metal to *Channa punctatus*. *J. Environ. Biol.*, 4(2): 91–94.

Sunderman, F.W. Jr., Dingle, B., Hopfer, S.M. and T. Swift (1988a), Acute nickel toxicity in electroplating workers who accidentally ingested a solution of nickel sulphate and nickel chloride. *Amer. J. Ind. Med.*, 14; 257–266.

Chapter 23

Removal of Heavy Metals from Electroplating Industrial Effluent Using Plants–Phytoremediation

S. Anitha, V. Mahesh and C. Sheela Sasikumar

ABSTRACT

Toxic metal contamination of soil, aqueous waste streams and groundwater causes major environmental and human health problems. The most commonly used methods for dealing with heavy metal pollution are still extremely costly. Phytoremediation is emerging as an innovative tool with greater potential for achieving sustainable development to decontaminate metal polluted air, soil and water. Plants are the ideal agents for soil and water remediation because of their unique genetic, biochemical and physiological properties. The present study is an attempt to investigate the ability of easily available and inexpensive plant bark to adsorb heavy metals from electroplating industrial effluent.

Keywords: *Adsorption, Heavy metal, Plant bark, Electroplating effluent, Phytoremediation.*

Introduction

The current concern for increased soil contamination due to both industrial and domestic pollution depends on several economic and health considerations: in many cases contaminated sites cannot be safely reused unless prior remediation measures are implemented (Hagemeyer, 1999), long-term effects may transform toxic compounds into non-predictable and even more toxic hazardous substances that may leak into water sources; and the volume of domestic and industrial waste is increasing continuously (Ghosh *et al.*, 2000).

Heavy metals are highly toxic because of their non-biodegradable nature and is found generated especially 90 per cent in electroplating industries alone (Longman, 1996). Plants can be compared to solar driven pumps, that extract and concentrate the elements from the environment (atleast 5 per cent of its dry weight). All plants extract necessary nutrients, including metals, from their soil and water environments (Mofa, 1995). Some plants, called hyperaccumulators, have the ability to store large amounts of metals, even some metals that do not appear to be required for plant functioning (Williams, 2000).

In the present study, an attempt has been made to record the phytoremediation capacity of *Tamarindus indicus* bark and *Pongamia glabra* bark. The influence of contact time on the adsorption of heavy metal like Cr, Ni, Zn, Pb in the electroplating effluent has been investigated (Singh *et al.*, 1994: Martin-Dupont *et al.*, 2002).

Materials and Methods

Tamarind and pongamia barks collected were dried in room temperature and ground. 500mg of bark powder was taken in conical flasks containing 100ml of effluent. The setup was agitated in a mechanical shaker for 30, 60, 90, 120 minutes at room temperature. Estimation of heavy metals were done following the standard procedures (Basett *et al.*, 1986).

Results and Discussion

Plants that hyper accumulate metals have tremendous potential for application in remediation of metals in the environment (Azadpour *et al.*, 1996). The control values were found to be 50,40 15 and 150 mg for Zn, Ni, Pb, Cr respectively.

Pongamia barks exhibited some appreciable affinity for the binding of Pb ions from the solution by 22 per cent to 56 per cent during the treatment. It is observed during the investigation that Cr concentration has decreased from 150 mg to 112.5 mg level in 60 min and the per cent reduction is 25. It can also be seen from the graph that there is a further decline in Cr steadily with increase in contact time up to 120 min. (40 per cent).

Table 23.1: Effect of Tamarind (*Tamarindus indicus*) Bark on Heavy Metal Concentration in Electroplating Effluent

Time	*Heavy Metals (mg/l)*			
	Zinc	*Nickel*	*Lead*	*Chromium*
0	50	40	15	150
30	31.66	35	13.33	130
60	29.99	35	11.66	112.5
90	25	30	9.99	105
120	23.33	30	9.99	101

Table 23.2: Effect of Pongamia (*Pongamia glabra*) Bark on Heavy Metal Concentration in Electroplating Effluent

Time	*Heavy Metals (mg/l)*			
	Zinc	*Nickel*	*Lead*	*Chromium*
0	50	40	15	150
30	33.33	35	11.66	120
60	16.66	30	9.99	112.5
90	13.33	30	8.33	100
120	11.66	25	6.66	97.5

It is interesting to note that removal efficiency of Zn by tamarind bark is higher at 30 min. time (37 per cent) compared to Pongamia (33 per cent). But as time passes (120 min) removal efficiency of Pongamia is more (77 per cent). The graph shows the faster adsorption of Ni ions from the effluent especially in first 30 min. (13 per cent). Whereas Pongamia could adsorb only 38 per cent even after 120 min. Plants can take up heavy metals by their roots, or even via their stems and leaves, and accumulate them in their organs (Cheng, 2003).

The metal uptake capacity of the Pongamia bark was found to be in the order of: Zn > Pb > Cr > Ni. The metal uptake capacity of the Tamarind bark was found to be in the order of: Zn > Cr > Ni > Pb. The treatment proved very economical and versatile, and the product

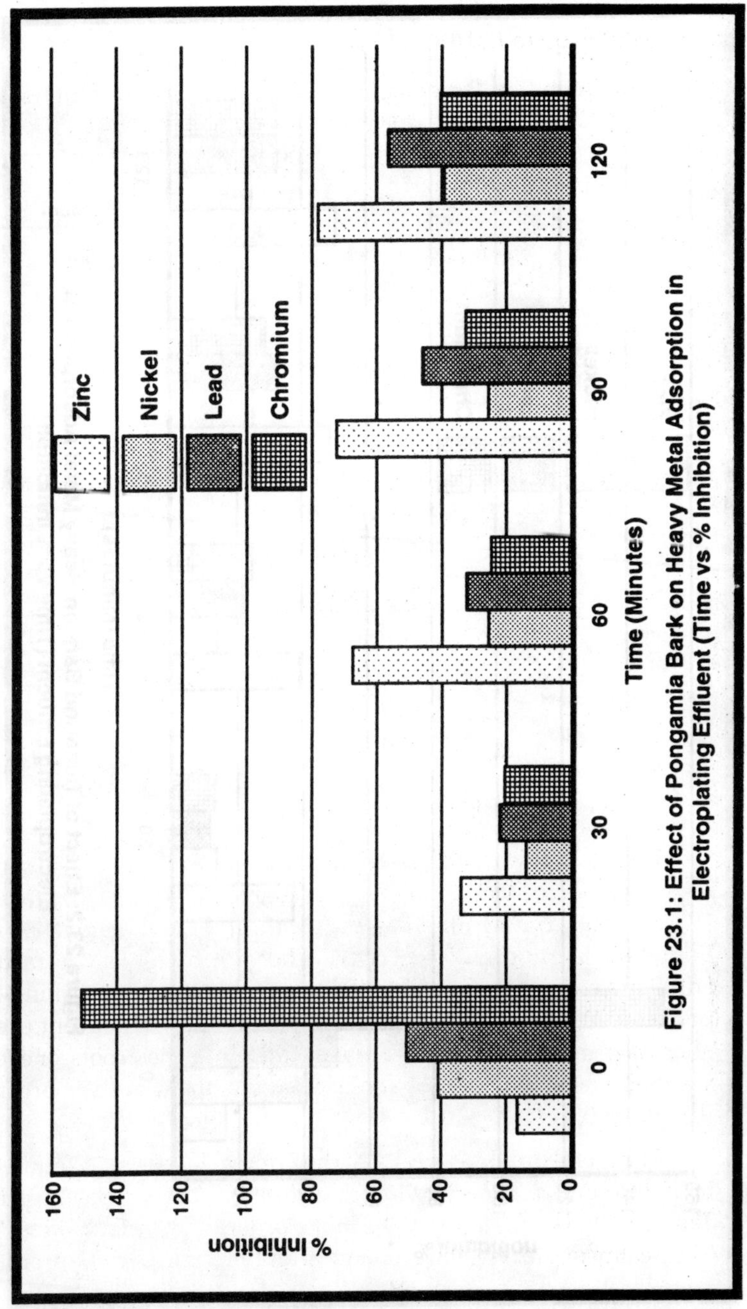

Figure 23.1: Effect of Pongamia Bark on Heavy Metal Adsorption in Electroplating Effluent (Time vs % Inhibition)

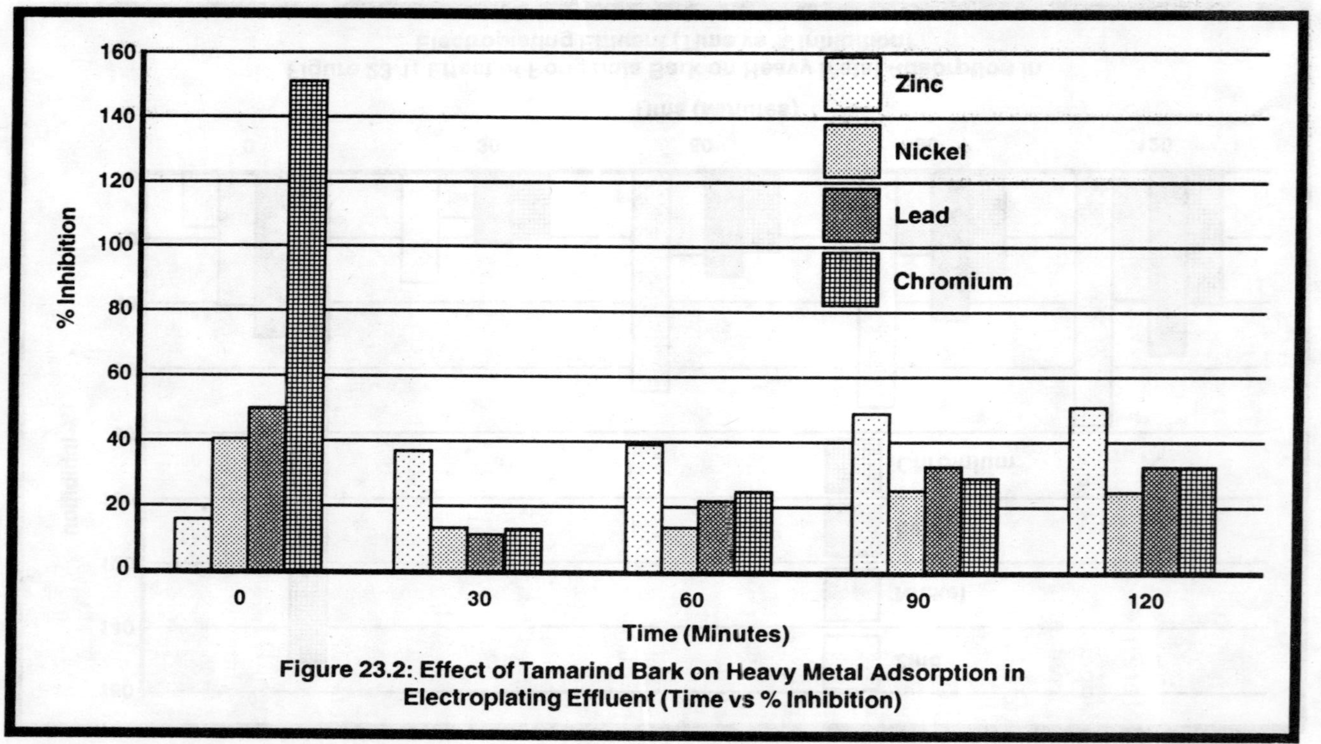

Figure 23.2: Effect of Tamarind Bark on Heavy Metal Adsorption in Electroplating Effluent (Time vs % Inhibition)

is easy to work with and safe to handle. The new treatment produces effluents that meet or exceed discharge standards, and generates non-hazardous waste. Importantly, the facility has realized a 10 per cent annual cost saving for its wastewater treatment.

References

Azadpour, A. and J.E. Matthews (1996). Remediation of metal-contaminated sites using plants. *Remed. Summer.* 5(3): 1–19.

Basett, J., Denney, R.C. and G.H. Jeffery (1986). *Vogel's Textbook of Quantitative Inorganic Analysis.* Calorimetry and Spectrophotometry Chapter XVIII, p. 738–772.

Cheng, S. (2003). Heavy metals in plants and phytoremediation. *Env. Sci. Pollut. Res. Int.*, 10(5): 335–340.

Ghosh, P.B., Saha, T., Bandyopadhaya, T.S. (2000). Distribution of lead, cadmium and chromium in the wastewater of Kolkata canals. *Res. J. Chem. Env.*, 4(3): 33–36.

Hagemeyer, J. (1999). Ecophysiology of plant growth under heavy metal stress. In: *Heavy Metal Stress in Plants,* (Eds.) M.N.V. Prasad and J. Hagemeyer. Springer-Verlag, Berlin, Heidelberg, p. 157–181.

Longman, Harlow, Charanjit Singh, Parwan, H.K., Merwaha, S.S., Garg, Rakesh and Gajendra Singh (1996). Toxicity of electroplating effluents. *Poll. Res.*, 12(1): 15–19.

Martin-Dupont, F., Gloaguen, V., Granet, R., Guilloton, M., Morvan, H. and P. Krausz (2002). Heavy metal adsorption by crude coniferous barks: A modelling study. *J. Env. Sci Health Part A: Tox Hazard Subst Environ Eng.*, 37(6): 1063–1073.

Mofa, A.S. (1995). Plants proving their worth in toxic metal cleanup. *Science*, 269: 302–305.

Singh, D.K., Saksena, D.N. and D.P. Tiwari (1994). Removal of chromium (VI) from aqueous solutions. *Indian J. Env. Hlth.*, 36(4): 272–277 (17Ref).

Williams, L.E., Pittman, J.K. and J.L. Hall (2000). Emerging mechanisms for heavy metal transport in plants. *Biochem. Biophys. Acta,* 1465(12): 104–126.

Chapter 24

Concentrations of Heavy Metals in *Penaeus* spp. of Brackishwater Wetland Ecosystem of West Bengal, India

A. Mitra, T. Mandal and D.P. Bhattacharyya

ABSTRACT

Concentrations of Zn, Cu, Mn, Fe, Cr, Ni, Pb and Cd were analysed monthly for two years in different body parts (muscle, gill, hepatopancreas and exoskeleton) of edible prawns, Penaeus monodon and Penaeus indicus sampled from Kulti brackishwater wetland system. The area receives the municipal wastes from the highly urbanised and industrialised city of Kolkata, which brings considerable amount of conservative wastes in the wetland system. Shrimp culture with the wastewater is a common practice in this area and therefore the probability of bioaccumulation of conservative wastes in the prawn tissues cannot be ignored. The concentrations of heavy metals in the prawn tissues exhibited unique seasonal variations and significant negative relationships with the aquatic pH and salinity of the study area.

Keywords: *Conservative wastes, Heavy metals, Seasonal variations, Aquatic pH, Aquatic salinity.*

Introduction

Edible Prawns, *Penaeus monodon* and *Penaeus indicus* are among the list of exportable items of the Indian Sub-continent which have unique potentiality to earn foreign exchange. However due to their culture in some wrong sites, the quality of the cultured products deteriorates to a great extent making the processing technology more expensive and difficult. Based on this problem, the present article aims to highlight the magnitude of deterioration of edible prawn tissues due to accumulation of heavy metals Zn, Cu, Mn, Fe, Cr, Ni, Pb and Cd sampled from the brackishwater pond of Kulti region, 35 km South East from the city of Kolkata. The area receives all the waste materials that arise due to intense industrial and human activities from the city of Kolkata.

The influence of physico-chemical variables (like heavy metals of the ambient media, surface water temperature, pH and salinity) on the metal concentrations in the prawn tissues have also been assessed in the programme as these variables have considerable effect on the process of uptake of heavy metals from the ambient media.

Materials and Methods

The entire network comprised the monthly sampling of same sized *Penaeus monodon* and *Penaeus indicus* from the brackishwater body of Kulti region during 1992 to 1994, along with water and surface sediments. Physico-chemical variables like surface water temperature, pH and salinity were also analysed simultaneously to determine their effect on the process of uptake of heavy metals by the prawns.

The heavy metals Zn, Cu, Mn, Fe, Cr, Ni, Pb and Cd were analysed in the sediment and water through Atomic Absorption Spectrophotometer (Perkin Elmer Type, Model 3030) as per the procedures outlined by Malo (1977) and Chakraborty *et al.* (1987), respectively. Both the prawn species were also carefully dissected to segregate muscles, hepatopancreas, gill and exoskeleton to determine concentrations of heavy metals in them through the same AAS after the process of acid digestion (Harper *et al.*, 1989).

The results obtained were then analysed statistically using SYSTAT Programme (Wilkinson, 1992) to analyse the influence of physico-chemical variables on heavy metal concentrations in the prawn tissues.

Results and Discussion

The physico-chemical variables showed unique seasonal variations in the sampling station. Surface water temperature was high during the premonsoon (March to June) and monsoon seasons (July to October) and low in the post monsoon months (November to February). The surface water salinity and pH were highest during the premonsoon months and lowest during the monsoon season (Table 24.1) which might be due to heavy precipitation and subsequent discharge of fresh water (run off) from the adjacent city of Kolkata during the monsoon. Similar observations were recorded by Mitra and Choudhury (1993) while working in the Hooghly estuarine region of the state of West Bengal.

The concentrations of heavy metals in the ambient media (sediment and water) also exhibited a sharp seasonal oscillation. The dissolved heavy metals were found to be maximum during the monsoon period, the period characterised by extremely low salinity and pH of the ambient aquatic medium and during premonsoon period, the concentrations of dissolved heavy metals reached to a minimum value (Table 24.2). This may be the effect of water salinity and pH which play a significant role in the process of compartmentation of heavy metals in the estuarine system of the West Bengal coast (Mitra *et al.*, 1994). The lowering of water pH and salinity due to increased precipitation switches on the process of dissolution of metallic compounds from the sediment compartment to the aquatic column (Lakshmanan and Nambisan, 1983) resulting in the increase of the dissolved heavy metals and decrease of the biologically available heavy metals in the sediment compartment during monsoon (Table 24.3). High concentrations of dissolved heavy metals during monsoon period have been recorded very recently in the coastal zone of West Bengal and in the present situation there is a keen relationship of the Kolkata run off that brings huge load of heavy metals during monsoon from several lead factories, electroplating units and tanneries in the brackishwater wetlands of the study area (Mitra, 1998).

The tissue accumulation of heavy metals were more in *Penaeus monodon* in comparison to *Penaeus indicus* although both the species inhabited the same water body. This clearly reflects more credibility of the later species in respect of membrane permeability.

The tissue distribution of the heavy metals in the edible prawns as a result of bioaccumulation, showed differences and in some cases high degree of organ specificity. In both the species Zn, Mn, Fe and Ni accumulated in the body tissues in the order gill > hepatopancreas > muscle > exoskeleton. In case of Cr, Cd and Pb the tissue accumulation is in the order gill > hepatopancreas > exoskeleton > muscle. In case of Cu, the order of accumulation in the body tissues is hepatopancreas > gill > muscle > exoskeleton in both the prawn species.

A unique seasonal variations was observed in case of heavy metal accumulation in the body tissues of both the edible prawn species. Accumulation was found to be maximum during rainy season and minimum during the premonsoon period in all types of body tissues (Tables 24.4 a, b, c, d and 24.5 a, b, c, d). This might be attributed to maximum run-off that occurs during the monsoon period and brings huge quantum of metallic wastes from the highly urbanised and industrialised city of Kolkata. The increment of dissolved metal concentrations in the ambient aquatic phase during monsoon period and their unique positive correlations with the tissue concentrations ($p < 0.01$) strongly support the view. The second cause behind high concentration of heavy metals in the body tissues during monsoon period may be related to the lowering of salinity and pH which might facilitate the dissolution of precipitated form of metal and increase the amount of metallic ions in solution (Lakshmanan and Nambisan, 1983). These two causes are supported by the significant negative correlations of tissue metal concentrations with salinity and pH of the aquatic medium of the selected sampling station and significant positive correlation between dissolved metal concentrations and respective tissue metals (Tables 24.6).

The highly significant negative correlations of aquatic pH and heavy metal concentrations of edible prawn tissues of the study area suggests the addition of calculated amount of lime in the brackishwater bodies to decrease the availability of metallic ions to the prawn species from the ambient aquatic phase.

Table 24.1: Monthly Variations of Physico-chemical Variables in Brackishwater Body of Kulti Station

Month	*Surface Water Temperature (°C)*	*pH*	*Salinity (ppt.)*
Mar 1992	29.8	8.30	16.93
Apr	30.6	8.31	17.10
May	31.3	8.30	18.21
Jun	32.8	8.31	19.36
Jul	30.2	8.00	8.83
Aug	31.5	7.99	7.08
Sep	30.4	7.68	5.66
Oct	29.3	7.74	5.14
Nov	28.6	7.91	9.13
Dec	26.2	7.99	9.71
Jan 1993	25.4	8.01	10.15
Feb	27.8	8.22	10.47
Mar	29.3	8.29	15.89
Apr	31.0	8.31	17.28
May	31.5	8.32	18.01
Jun	33.0	8.33	18.86
Jul	31.6	8.06	8.14
Aug	30.5	8.07	6.59
Sep	30.8	7.91	5.47
Oct	30.0	7.69	5.02
Nov	29.1	7.97	8.43
Dec	27.6	8.21	8.76
Jan 1994	26.8	8.26	9.34
Feb	27.1	8.30	9.85

Table 24.2: Monthly Variations of Dissolved Heavy Metals (in μg l^{-1}) in Brackishwater Body of Kulti Station

Month	*Zn*	*Cu*	*Mn*	*Fe*	*Cr*	*Ni*	*Pb*	*Cd*
Mar 1992	212.7	182.4	315.3	3062.3	7.4	25.2	70.4	5.2
Apr	217.4	189.7	327.5	3095.7	8.1	27.4	73.8	5.7
May	225.3	197.8	332.8	3104.1	9.3	29.2	76.2	5.9
Jun	232.5	201.3	340.2	3163.2	10.2	30.4	78.1	6.1
Jul	279.1	245.2	363.1	4520.3	23.1	35.6	89.7	8.3
Aug	287.0	264.7	372.3	4617.4	26.2	38.2	97.5	9.5
Sep	295.6	279.1	385.4	4793.5	29.5	39.3	104.6	10.1
Oct	312.3	285,3	391.3	4872.6	32.6	41.7	118.4	11.4
Nov	292.1	263.1	372.1	3570.4	25,1.	36.2	90.3	9.2
Dec	280.4	251.4	360.4	3645.5	21.4	34.1	82.4	8.7
Jan 1993	273.5	243.3	349.0	3802.4	20.3	32.4	77.6	8.1
Feb	262.4	232.1	342.9	3917.2	18.2	30.7	74.3	7.5
Mar	202.3	170.4	303.4	3042.4	7.1	23.5	67.2	5.1
Apr	206.5	176.7	311.2	3034.2	7.8	25.2	69.4	5.4
May	214.2	184.3	319.4	3097.5	9.2	26.5	73.5	5.5
Jun	220.4	190.4	326.7	3157.4	10.0	28.2	75.2	6.0
Jul	264.3	230.3	338.4	4564.6	22.7	32.9	86.3	7.9
Aug	271.6	251.6	361.5	4634.7	25.3	35.7	94.2	8.9
Sep	280.9	267.5	370.3	4809.4	27.4	37.3	100.3	9.7
Oct	299.7	272.4	381.4	4889.7	30.2	39.2	108.5	10.2
Nov	282.3	248.3	360.2	3554.3	24.6	35.7	86.2	8.9
Dec	269.5	239.4	349.4	3637.2	19.3	30.2	80.3	8.0
Jan 1994	262.7	230.6	331.3	3781.8	18.2	29.1	71.4	7.7
Feb	251.3	221.2	330.5	3900.7	17.7	28.3	70.7	7.1

Table 24.3: Monthly Variations of Biologically Available Heavy Metals (in mg Kg^{-1} dry wt.) in the Pond Bottom Sediment of Kulti Brackishwater Body

Month	*Zn*	*Cu*	*Mn*	*Fe*	*Cr*	*Ni*	*Pb*	*Cd*
Mar 1992	221.10	97.80	110.60	10002.40	40.10	13.60	20.20	2.40
Apr	213.20	88.70	126.30	9443.60	37.60	14.80	16.60	2.30
May	212.20	74.30	81.50	10064.70	36.40	16.60	18.40	2.50
Jun	220.50	102.20	94.30	9788.60	34.20	22.30	12.10	2.70
Jul	182.20	71.40	66.30	7506.70	27.70	5.20	12.40	–
Aug	165.40	38.70	84.10	7649.70	26.30	4.40	11.70	–
Sep	159.50	46.80	104.40	6333.90	25.10	3.90	10.20	–
Oct	172.70	53.20	93.30	6760.30	21.90	3.80	16.70	–
Nov	191,60	65.40	101.40	9542.40	26.30	6.50	16.30	–
Dec	184.10	57.70	99.60	9697.20	26.80	8.30	19.10	–
Jan 1993	190.30	92.50	78.69	8029.20	10.50	17.80	–	–
Feb	199.40	51.40	100.30	6557.30	30.40	10.90	16.20	2.60
Mar	118.30	84.30	91.20	11665.50	39.30	13.30	19.80	2.40
Apr	216.50	82.70	75.90	8212.60	38.70	17.10	12.40	2.30
May	221.60	79.40	73.10	6549.10	36.60	29.40	12.70	2.20
Jun	218.30	85.60	86.30	10024.90	37.40	16.80	14.20	2.40
Jul	158.50	69.20.	87.20	8810.60	28.30	7.30	12.30	–
Aug	166.30	59.50	96.70	7635.30	26.10	4.20	11.60	–
Sep	158.40	63.30	103.50	8736.80	24.50	3.10	8.80	–
Oct	174.30	59.20	91.80	7941.50	22.30	6.30	7.90	–
Nov	188.50	71.40	100.80	10812.70	26.20	7.30	9.30	–
Dec	201.60	64.80	91.10	8051.80	27.40	8.80	13.10	–
Jan 1994	205.30	90.50	77.47	8029.40	10.80	16.40	–	–
Feb	211.40	55.70	103.30	6916.20	30.10	13.30	19.30	2.30

–: means below detectable level.

Table 24.4a: Mean Monthly Heavy Metal Concentrations (in µg gm^{-1} dry wt.) in the Muscles of *Penaeus monodon* Sampled from Kulti Brackishwater Body

Month	*Zn*	*Cu*	*Mn*	*Fe*	*Cr*	*Ni*	*Pb*	*Cd*
Mar'92	103.45	16.34	20.21	164.04	–	–	2.81	–
Apr	101.86	17.87	19.34	159.24	–	–	2.56	–
May	110.88	14.32	18.73	154.93	–	–	–	–
Jun	100.34	12.64	23.45	144.93	–	–	2.17	–
Jul	121.64	20.53	26.24	168.38	–	2.01	2.98	–
Aug	130.85	21.65	33.17	207.83	2.20	2.31	3.09	–
Sept	134.36	24.78	37.26	220.11	2.26	5.06	6.68	2.18
Oct	139.38	26.84	38.32	217.05	–	4.86	5.91	–
Nov	122.21	18.21	30.26	212.86	–	3.32	5.03	–
Dec	126.20	22.08	33.18	209.52	–	–	3.94	–
Jan'93	117.08	20.83	29.25	177.91	–	–	3.83	–
Feb	108.63	19.77	26.42	171.09	–	–	3.83	–
Mar	111.72	21.63	26.76	172.50	–	–	2.98	–
Apr	107.29	22.54	29.17	167.06	–	–	2.86	–
May	115.27	18.17	27.63	148.73	–	–	2.01	–
Jun	112.93	17.68	31.25	138.77	–	–	2,67	–
Jul	120.84	26.71	34.17	160.47	–	3.68	5.73	–
Aug	136.32	25.03	37.41	217.05	2.43	4.75	6.20	2.25
Sept	144.85	27.63	36.32	235.02	2.18	4.32	7.85	–
Oct	128.71	23.29	28.55	213.14	–	3.96	6.53	–
Nov	123.25	19.57	30.46	196.33	–	2.78	5.81	–
Dec	110.64	16.82	20.81	205.16	–	–	5.37	–
Jan'94	117.56	14.35	26.19	183.39	–	–	3.08	–
Feb	113.21	18.29	22.20	170.03	–	–	3.63	–

–: means below detectable level.

Table 24.4b: Mean Monthly Heavy Metal Concentrations (in µg gm⁻¹ dry wt.) in the Gills of *Penaeus monodon* Sampled from Kulti Brackishwater Body

Month	*Zn*	*Cu*	*Mn*	*Fe*	*Cr*	*Ni*	*Pb*	*Cd*
Mar'92	270.21	123.36	65.23	592.17	4.69	3.68	7.11	4.25
Apr	231.45	111.45	61.08	576.29	4.85	2.33	5.44	4.08
May	202.83	109.63	55.73	538.00	4.13	2.01	5.00	3.32
Jun	212.67	99.47	43.96	502.89	2.98	–	3.71	2.90
Jul	277.83	142.75	76.83	613.16	7.93	5.13	8.07	6.25
Aug	293.37	158.21	91.35	653.22	11.66	9.71	12.00	10.21
Sept	307.20	169.35	87.10	671.85	12.01	9.80	13.17	10.66
Oct	301.00	151.80	86.93	603.98	11.00	8.55	12.85	9.09
Nov	278.65	144.86	60.55	615.00	10.96	7.23	11.28	8.10
Dec	283.11	137.13	78.31	591.87	10.11	6.20	11.07	7.16
Jan'93	281.93	131.67	76.87	543.06	6.83	3.99	9.80	4.23
Feb	274.05	129.84	71.08	536.78	4.99	3.21	8.116	4.52
Mar	281.93	127.71	70.71	568.11	5.38	3.83	8.77	4.00
Apr	237.76	109.56	68.27	556.30	5.92	2.71	7.88	4.84
May	212.24	100.29	62.62	523.57	4.08	3.09	4.66	3.20
Jun	201.58	113.85	50.54	519.32	4.00	3.11	4.01	3.96
Jul	287.37	150.27	68.20	672.69	7.68	4.08	7.98	6.83
Aug	296.28	168.68	79.00	638.98	9.38	6.83	11.66	10.87
Sept	309.71	170.53	83.37	681.00	15.29	7.99	14.84	12.65
Oct	299.82	149.49	80.21	616.59	13.26	7.84	12.37	10.85
Nov	280.65	158.75	76.52	598.27	10.85	6.37	10.65	9.33
Dec	280.03	207.29	79.99	578.44	7.21	4.89	9.32	6.88
Jan'94	426.55	122.31	71.81	537.12	6.88	4.00	8.44	5.10
Feb	248.00	128.68	74.26	592.33	5.81	2.31	7.93	4.68

–: means below detectable level.

Table 24.4c: Mean Monthly Heavy Metal Concentrations (in µg gm^{-1} dry wt.) in the Hepatopancreas of *Penaeus monodon* Sampled from Kulti Brackishwater Body

Month	*Zn*	*Cu*	*Mn*	*Fe*	*Cr*	*Ni*	*Pb*	*Cd*
Mar'92	144.31	161.63	38.31	411.41	4.11	4.03	3.38	2.07
Apr	157.81	159.57	31.99	408.88	4.01	3.89	2.28	2.18
May	148.39	155.39	31.99	400.09	4.08	3.21	2.87	–
Jun	143.97	149.73	28.81	396.22	2.42	3.24	2.89	2.73
Jul	209.73	281.94	60.01	483.13	5.41	5.33	4.08	3.01
Aug	211.51	229.33	68.93	514.71	9.98	9.92	6.21	4.99
Sept	264.78	283.57	71.99	585.87	11.00	10.58	8.92	5.63
Oct	215.31	242.52	69.23	547.17	10.91	10.87	9.63	9.00
Nov	179.13	183.09	66.91	526.35	8.06	7.81	6.37	6.05
Dec	170.79	181.91	58.01	496.99	9.20	6.92	5.07	4.00
Jan'93	160.01	177.53	58.23	427.01	6.80	4.10	3.38	3.18
Feb	143.23	168.27	43.05	417.65	4.62	3.89	2.90	2.40
Mar	146.56	164.53	39.39	414.37	9.15	4.07	3.41	–
Apr	149.43	162.37	34.72	411.72	4.05	3.93	2.15	–
May	150.93	158.12	32.84	403.83	4.12	3.75	2.90	2.31
Jun	145.17	152.37	29.31	302.14	2.46	2.25	2.03	–
Jul	212.39	214.38	62.17	487.33	5.45	5.37	4.10	4.04
Aug	214.27	242.48	70.43	518.34	10.02	9.96	6.24	5.02
Sept	267.35	286.73	73.41	588.57	11.04	10.62	8.95	7.66
Oct	218.18	265.27	70.51	551.65	8.95	8.91	7.66	5.98
Nov	181.17	185.45	67.67	529.53	8.10	7.85	6.40	4.08
Dec	172.39	182.39	60.28	499.15	9.24	6.96	5.10	3.03
Jan'94	162.17	179.34	30.38	431.34	6.84	4.14	3.41	2.21
Feb	146.37	170.71	45.14	421.57	4.66	3.93	2.93	2.03

–: means below detectable level.

Table 24.4d: Mean Monthly Heavy Metal Concentrations (in µg gm⁻¹ dry wt.) in the Exoskeleton of *Penaeus monodon* Sampled from Kulti Brackishwater Body

Month	*Zn*	*Cu*	*Mn*	*Fe*	*Cr*	*Ni*	*Pb*	*Cd*
Mar'92	93.70	8.83	10.35	114.35	2.88	–	2.97	–
Apr	91.65	7.69	10.57	110.97	–	–	2.68	–
May	83.28	7.47	9.88	107.66	–	–	2.85	–
Jun	78.62	5.98	8.09	101.00	–	–	2.32	–
Jul	99.07	9.37	11.67	117.34	–	–	3.01	–
Aug	101.33	16.23	19.28	144.83	3.01	–	3.67	2.93
Sept	108.84	17.01	21.13	153.39	2.66	2.01	7.00	2.01
Oct	100.65	14.23	16.56	151.26	3.98	2.03	6.23	2.67
Nov	97.73	12.89	14.88	148.34	3.19	–	5.19	2.03
Dec	93.88	11.33	13.29	146.01	2.83	–	4.86	–
Jan'93	89.27	11.00	12.61	123.98	–	–	3.85	–
Feb	90.26	9.58	10.01	119.23	–	–	3.23	–
Mar	97.12	9.07	11.66	120.21	4.00	–	4.06	–
Apr	103.44	111.42	19.92	341.74	–	–	–	–
May	86.82	5.98	10.50	103.65	–	–	2.98	–
Jun	74.21	5.08	10.26	96.71	–	–	2.91	–
Jul	104.35	10.77	14.85	111.83	6.64	–	7.63	–
Aug	111.00	13.85	21.36	151.26	7.01	–	7.70	–
Sept	117.03	19.23	18.93	163.78	8.33	–	9.17	3.46
Oct	96.47	15.67	17.65	148.53	–	–	6.67	–
Nov	81.65	13.82	13.29	136.82	–	–	6.08	–
Dec	105.49	11.27	12.85	142.97	–	–	5.71	–
Jan'94	79.38	10.85	10.89	121.80	–	–	4.81	–
Feb	88.71	9.39	11.00	118.49	4.27	–	4.21	–

–: means below detectable level.

Table 24.5a: Mean Monthly Heavy Metal Concentrations (in µg gm^{-1} dry wt.) in the Muscles of *Penaeus indicus* Sampled from Kulti Brackishwater Body

Month	*Zn*	*Cu*	*Mn*	*Fe*	*Cr*	*Ni*	*Pb*	*Cd*
Mar'92	98.38	13.49	15.47	144.51	–	–	–	–
Apr	94.43	15.53	16.32	137.46	–	–	–	–
May	103.71	10.64	14.46	134.63	–	–	–	–
Jun	99.54	8.72	18.51	123.71	–	–	–	–
Jul	112.65	16.48	20.28	149.34	3.17	4.25	3.19	–
Aug	122.42	18.53	25.34	188.52	3.43	4.42	3.58	–
Sept	125.67	20.64	30.21	203.41	4.03	4.78	4.52	2.48
Oct	127.53	21.71	32.51	200.51	–	5.47	5.31	3.17
Nov	111.34	15.82	24.46	190.63	–	4.33	4.17	3.29
Dec	117.21	14.48	25.19	183.75	–	–	–	–
Jan'93	110.47	13.57	22.24	156.34	–	–	–	–
Feb	99.71	12.45	20.52	153.47	–	–	–	–
Mar	97.52	12.98	16.34	145.63	–	–	–	–
Apr	102.34	14.43	17.47	140.71	–	–	–	–
May	104.48	12.18	15.32	137.28	–	–	–	–
Jun	106.72	9.23	18.17	127.39	–	–	–	–
Jul	116.81	16.84	19.32	153.25	3.29	4.31	3.38	–
Aug	125.34	19.31	26.71	192.32	3.91	4.22	3.53	–
Sept	127.17	19.98	31.25	208.41	4.27	4.82	4.71	2.51
Oct	118.32	22.17	34.34	207.23	–	5.64	5.52	3.45
Nov	115.45	16.25	23.73	193.17	–	4.68	4.10	3.01
Dec	106.43	15.34	25.34	187.63	–	–	–	–
Jan'94	111.37	14.63	21.42	155.71	–	–	–	–
Feb	100.32	12.81	21.74	151.21	–	–	–	–

–: means below detectable level.

Table 24.5b: Mean Monthly Heavy Metal Concentrations (in µg gm^{-1} dry wt.) in the Gills of *Penaeus indicus* Sampled from Kulti Brackishwater Body

Month	*Zn*	*Cu*	*Mn*	*Fe*	*Cr*	*Ni*	*Pb*	*Cd*
Mar'92	251.72	114.62	59.71	551.37	5.04	4.17	5.12	3.17
Apr	208.45	103.27	54.84	532.42	5.27	4.62	4.77	3.05
May	177.64	100.48	47.32	498.51	5.56	7.74	4.88	–
Jun	184.53	87.24	40.17	457.75	5.97	4.98	–	–
Jul	262.42	138.14	71.34	567.82	6.61	5.72	6.32	5.17
Aug	269.74	148.27	83.25	612.63	10.72	9.31	10.17	9.25
Sept	285.65	159.18	79.17	625.74	12.34	10.67	10.93	9.47
Oct	281.57	140.34	77.25	562.63	10.53	10.45	12.24	8.32
Nov	213.25	136.12	74.37	571.54	9.74	8.95	10.18	7.15
Dec	248.18	131.24	70.38	552.82	8.12	6.37	10.05	6.14
Jan'93	258.26	122.72	69.33	503.69	6.14	5.31	7.32	3.24
Feb	252.57	120.34	61.48	592.71	6.01	4.18	6.14	3.47
Mar	256.34	111.39	62.38	554.84	5.19	4.03	6.73	3.13
Apr	213.72	100.48	56.72	536.36	5.38	3.94	5.42	3.57
May	184.17	103.27	48.48	501.45	5.67	4.28	–	
Jun	180.22	85.85	43.75	455.39	5.83	4.37	–	
Jul	271.41	142.45	70.17	572.54	8.18	4.51	6.14	4.21
Aug	280.25	151.71	85.36	615.27	13.27	8.45	10.25	8.32
Sept	291.54	158.53	82.47	626.45	11.98	9.76	11.03	9.56
Oct	287.42	143.24	75.55	565.17	9.78	8.40	12.74	8.33
Nov	232.61	140.41	73.63	576.64	9.04	7.31	10.79	7.24
Dec	250.49	135.27	71.45	561.82	8.35	6.53	10.52	6.27
Jan'94.	269.17	125.38	68.56	509.65	6.17	5.19	7.15	3.56
Feb	250.24	122.54	60.86	499.47	5.65	5.32	6.37	3.69

–: means below detectable level.

Table 24.5c: Mean Monthly Heavy Metal Concentrations (in µg gm^{-1} dry wt.) in the Hepatopancreas of *Penaeus indicus* Sampled from Kulti Brackishwater Body

Month	*Zn*	*Cu*	*Mn*	*Fe*	*Cr*	*Ni*	*Pb*	*Cd*
Mar'92	114.27	136.56	32.42	363.32	2.11	2.17	5.18	2.08
Apr	129.49	132.27	26.77	356.73	2.09	2.84	4.17	2.03
May	119.36	128.34	25.82	356.39	2.34	2.97	4.37	–
Jun	111.53	121.45	22.33	343.24	2.12	2.97	4.11	2.01
Jul	186.64	261.63	53.39	431.45	3.37	3.56	6.13	3.24
Aug	189.67	208.62	61.46	463.47	7.28	7.92	8.24	5.47
Sept	241.71	262.67	65.56	532.53	9.14	9.85	10.15	6.71
Oct	191.55	221.73	62.37	497.63	6.37	9.83	10.92	5.14
Nov	151.37	150.81	59.57	473.71	6.02	6.43	7.34	5.13
Dec	141.45	158.93	51.13	441.81	4.85	5.74	6.17	3.43
Jan'93	129.33	152.87	52.34	371.07	3.57	5.01	5.97	3.27
Feb	109.42	142.34	37.51	367.57	2.64	3.97	4.12	2.15
Mar	115.39	139.31	33.24	364.83	2.36	2.65	4.53	–
Apr	132.67	135.49	28.63	355.34	2.73	2.93	3.55	–
May	121.51	131.32	26.31	351.97	2.94	3.02	3.47	2.07
Jun	112.64	122.51	23.72	165.48	2.51	3.03	3.12	–
Jul	188.32	266.31	55.37	434.39	3.03	3.72	5.14	3.01
Aug	190.37	211.47	63.29	465.73	6.13	7.14	7.43	6.04
Sept	244.32	265.32	67.27	535.52	8.74	9.96	10.17	8.57
Oct	194.46	223.49	63.17	499.62	7.97	10.01	10.12	6.77
Nov	155.53	161.77	61.04	475.81	5.63	5.74	7.53	5.43
Dec	144.59	160.83	53.24	442.73	5.03	5.53	6.24	4.92
Jan'94	131.29	155.33	53.77	374.33	4.29	3.53	4.53	3.37
Feb	112.37	146.27	39.11	369.64	2.73	3.04	3.73	2.14

–: means below detectable level.

Table 24.5d: Mean Monthly Heavy Metal Concentrations (in µg gm⁻¹ dry wt.) in the Exoskeleton of *Penaeus indicus* Sampled from Kulti Brackishwater Body

Month	*Zn*	*Cu*	*Mn*	*Fe*	*Cr*	*Ni*	*Pb*	*Cd*
Mar'92	74.26	6.93	8.27	105.36	3.57	–	4.01	2.17
Apr	71.32	5.75	8.68	89.47	2.26	–	3.92	2.16
May	63.73	5.68	7.99	87.51	2.18	–	3.98	2.39
Jun	57.17	4.97	7.01	81.32	2.03	–	3.55	2.03
Jul	79.32	7.12	9.76	106.41	2.81	2.65	4.17	–
Aug	84.18	13.55	17.72	122.53	3.92	2.81	5.68	3.72
Sept	85.76	14.67	18.51	130.64	3.84	2.17	6.73	3.67
Oct	83.38	12.51	14.63	128.71	4.07	–	7.18	3.69
Nov	74.25	10.17	12.87	125.34	3.63	–	5.79	3.07
Dec	71.34	9.35	11.04	123.19	3.79	–	5.19	2.52
Jan'93	66.54	9.86	10.63	103.47	2.47	–	4.27	2.32
Feb	68.69	7.14	8.36	98.81	2.53	–	4.03	2.37
Mar	76.73	7.15	8.63	107.34	3.74	–	4.81	2.52
Apr	70.39	6.25	8.71	95.46	2.42	–	4.72	2.37
May	65.81	5.91	8.05	86.17	2.31	–	4.89	2.21
Jun	61.34	5.16	7.63	78.09	2.27	–	5.71	2.14
Jul	75.73	7.71	10.07	111.34	3.09	2.42	5.22	3.01
Aug	85.55	14.01	16.63	125.72	4.05	2.73	5.82	3.81
Sept	84.77	15.09	17.69	137.17	3.98	2.91	6.93	3.65
Oct	84.18	12.79	15.97	129.34	4.19	2.36	7.72	3.76
Nov	76.32	9.87	13.32	122.19	3.52	–	6.07	2.83
Dec	74.83	9.92	11.65	120.37	3.87	–	5.58	2.41
Jan'94	67.15	10.71	11.01	109.71	2.74	–	4.17	2.34
Feb	70.44	7.95	8.74	101.83	2.63	–	4.18	2.27

–: means below detectable level.

Table 24.6: Inter-relationship Between Accumulated Heavy Metals in Prawn Tissues and Abiotic Variables. The sediment metal is actually the biologically available fraction of the total metal.

Combination	1 Muscle r-value	1 Gills r-value	1 Hepato-pancreas r-value	1 Exoskeleton r-value	2 Muscle r-value	2 Gills r-value	2 Hepato-pancreas r-value	2 Exoskeleton r-value
Tissue Zn × sediment Zn	–0.65**	–0.71**	–0.61**	–0.65**	–0.55*	–0.67**	–0.59*	–0.73**
Tissue Cu × sediment Cu	–0.54*	–0.61**	–0.56*	–0.71**	–0.60**	–0.74**	–0.53*	–0.72**
Tissue Mn × sediment Mn	0.09	–0.02	0.09	0.12	0.17	–0.13	0.08	0.15
Tissue Fe × sediment Fe	–0.26	–0.19	–0.29	–0.25	–0.21	–0.26	–0.24	–0.21
Tissue Cr × sediment Cr	–0.39	–0.83**	–0.84**	–0.25	–0.43*	–0.24	–0.22	–0.13
Tissue Ni × sediment Ni	–0.79**	–0.88**	–0.85**	–0.36	–0.80**	–0.72**	–0.78**	–0.64**
Tissue Pb × sediment Pb	–0.54*	–0.34	–0.56*	–0.51*	–0.62**	–0.26	–0.48*	–0.58*
Tissue Cd × sediment Cd	–0.25	–0.76**	–0.73**	–0.40**	–0.48*	–0.78**	–0.80**	–0.39
Tissue Zn × dissolved Zn	0.77**	0.77**	0.72**	0.50*	0.81**	0.65**	0.70**	0.61**
Tissue Cu × dissolved Cu	0.53*	0.72**	0.76**	0.87**	0.77**	0.86**	0.71**	0.84**
Tissue Mn × dissolved Mn	0.61**	0.62	0.85**	0.74**	0.88**	0.73**	0.82**	0.81**
Tissue Fe × dissolved Fe	0.69**	0.76**	0.78**	0.69**	0.73**	0.82**	0.80**	0.73**

Contd...

Table 24.6–Contd...

Combination	*1 Muscle r-value*	*1 Gills r-value*	*1 Hepato-pancreas r-value*	*1 Exoskeleton r-value*	*2 Muscle r-value*	*2 Gills r-value*	*2 Hepato-pancreas r-value*	*2 Exoskeleton r-value*
Tissue Cr × dissolved Cr	0.49*	0.86**	0.83**	0.36	0.53*	0.83**	0.81**	0.67**
Tissue Ni × dissolved Ni	0.84**	0.83**	0.86**	0.50*	0.87**	0.86**	0.89**	0.60**
Tissue Pb × dissolved Pb	0.75**	0.74**	0.85**	0.64**	0.91**	0.69**	0.93**	0.82**
Tissue Cd × dissolved Cd	0.29	0.86**	0.90**	0.62**	0.69**	0.86**	0.86**	0.57*
Tissue Zn × water temp.	–0.03	–0.38	0.15	–0.02	0.01	–0.36	0.17	–0.01
Tissue Cu × water temp.	–0.03	–0.22	0.09	–0.18	–0.06	–0.25	0.12	–0.20
Tissue Mn × water temp.	–0.09	–0.39	–0.17	0.05	–0.21	–0.33	–0.31	–0.02
Tissue Fe × water temp.	–0.32	0.01	–0.14	–0.31	–0.25	–0.07	–0.02	–0.31
Tissue Cr × water temp.	0.25	–0.12	–0.28	0.10	0.33	0.07	–0.04	–0.17
Tissue Ni × water temp.	0.21	–0.08	–0.02	0.02	0.21	0.03	–0.04	0.33
Tissue Pb × water temp.	–0.10	–0.33	–0.03	–0.07	0.18	0.42*	–0.05	0.11
Tissue Cd × water temp.	0.11	0.02	–0.09	0.14	–0.11	–0.19	–0.13	0.06
Tissue Zn × pH	–0.82**	–0.79**	–0.79**	–0.55*	–0.83**	–0.64**	–0.78**	–0.73**
Tissue Cu × pH	–0.66**	–0.60**	–0.79**	–0.83**	–0.84**	–0.78**	–0.74**	–0.77**

Contd...

Table 24.6–Contd...

Combination	*1 Muscle r-value*	*1 Gills r-value*	*1 Hepato-pancreas r-value*	*1 Exoskeleton r-value*	*2 Muscle r-value*	*2 Gills r-value*	*2 Hepato-pancreas r-value*	*2 Exoskeleton r-value*
Tissue Mn × pH	–0.67**	–0.67**	–0.89**	–0.75**	–0.84**	–0.78**	–0.86**	–0.79**
Tissue Fe × pH	–0.79**	–0.66**	–0.86	–0.79	–0.86	–0.61	–0.88	–0.82
Tissue Cr × pH	–0.34	–0.89**	0.81**	–0.26	–0.39	–0.77**	–0.83**	–0.71**
Tissue Ni × pH	–0.82**	–0.85**	–0.84**	–0.47*	–0.86**	–0.78**	–0.89**	–0.53*
Tissue Pb × pH	–0.77**	–0.84	–0.86**	–0.65**	–0.89**	–0.81**	–0.92**	–0.79**
Tissue Cd × pH	–0.22	–0.83**	–0.87**	–0.57*	–0.79**	–0.82**	–0.80**	–0.54*
Tissue Zn × salinity	–0.81**	–0.89**	–0.77**	–0.67**	–0.81**	–0.85**	–0.76**	–0.77*
Tissue Cu × salinity	–0.64**	–0.80**	–0.79**	–0.89**	–0.84**	–0.95**	–0.77**	–0.87**
Tissue Mn × salinity	–0.61**	0.84**	–0.89**	–0.78**	–0.87**	–0.83**	–0.96**	–0.93**
Tissue Fe × salinity	–0.84**	–0.74**	–0.87**	–0.84**	–0.88**	–0.73**	–0.85**	–0.90**
Tissue Cr × salinity	–0.46*	–0.85**	–0.86**	–0.39	–0.52*	–0.79**	–0.80**	–0.59*
Tissue Ni × salinity	–0.75**	–0.81**	–0.82**	–0.36	–0.76**	–0.69**	–0.80**	–0.65**
Tissue Pb × salinity	–0.81**	–0.88**	–0.80**	–0.77**	–0.76**	–0.87**	–0.80**	–0.65**
Tissue Cd × salinity	–0.31	–0.84**	–0.80**	–0.50*	–0.55*	–0.88**	–0.85**	–0.52*

1: *Penaeus monodon*, 2: *Penaeus indicus*; *: Significant at 5 per cent level; **: Significant at 1 per cent level.

References

Chakraborty, D., Adams, F., Van Mol, W. and J.K. Irgolic (1987). Determination of trace metals in natural waters at nanogram per litre levels by electrothermal atomic absorption spectrometry after extraction with sodium diethyldithio carbamate. *Analytica Chem. Acta*, 196: 23–31.

Harper, D.J., Fileman, C.F., May, P.V. and J.F. Postmann (1989). *Aquatic Environmental Protocols: Analytical Methods*. MAFF Direct Fish. Res. Lowestoff, U.K.

Lakshmanan, P.T. and P.N.K. Nambisan (1983). Seasonal variations in trace metal content in bivalve molluscs *Villorita cyprinoides, Meretrix casta* and *Perna viridis*. *Indian. J. Mar. Sci.*, 12: 100–103.

Malo, B.A. (1977). Partial extraction of metals from aquatic sediments. *Environ. Sci. Technol.*, 11: 217–288.

Mitra, Abhijit (1998). Status of coastal pollution in West Bengal with special reference to heavy metals. *J. Indian. Ocean Stud.*, 5(2): 135–138.

Mitra, Abhijit and Amalesh Choudhury (1993). Trace metals in macrobenthic molluscs of the Hooghly estuary, India. *Mar. Pollut. Bull.*, 26(9): 521–522.

Mitra, Abhijit, Trivedi, Subrata and Amalesh Choudhury (1994). Inter-relationship between trace metal pollution and physico-chemical variables in the frame work of Hooghly estuary. *Indian Ports*, p. 27–35.

Wilkinson, L. (1992). *SYSTAT for Windows*, version 5.02. SYSTAT, Evanston, IL.

Chapter 25

Chemical Impact on the Histological Studies on the Thyroid in the Freshwater Fish *Channa orientalis* (Sch.)

S.V. Deshmukh and K.M. Kulkarni

ABSTRACT

Thiourea treated thyroid gland showed sign of hyperactivity (increase in height of follicular epithelium). Hypofunction (non vacuolated colloid) and drastic atrophic changes (fibrosis and cyst formation).

Keywords: *Thyroid gland, Thiourea, Hyperactivity, Hypofunction.*

Introduction

Thiourea is an antithyroid drug, known to inhibit the synthesis of thyroxine. Due to its effects, animals can accumulate iodide in normal manner, but the iodination of thyroxine in the gland is blocked and thyroxine can not be synthesized. Thiourea made it easy to establish the functional relationship of thyroid gland with the pituitary gland and goands. This goitrogen compound provides a mean of chemical thyroidectomy useful in fishes. Since the surgical removal of thyroid gland is not possible because of the diffused structure of the gland, the method of chemical thyroidectomy by administrating thiourea has been employed in this study.

In fishes, the thyroid follicles lie scattered along the blood vessels under pharynx. In chondrichthyean fishes, some teleosts and lungfishes, the follicles of thyroid are arranged into distinct glandular mass. In some teleosts thyroid follicles may disperse from pharyngeal region to brain, kidney, eyes, spleen etc. In lampreys and bony fishes, the thyroid follicles tend to be dispersed along the ventral aorta.

The hormone that derived from thyroid gland is referred as thyroid hormone. The function of thyroid gland is to elaborate, store, and discharge secretions. Its basic function is to concentrate iodine and synthesize the thyroid hormone. The thyroid has greater capacity to store its secretions. Thyroid gland is composed cyst or oval like follicles of variable sizes. Each follicle is lined by single layer of cuboidal to low columnar epithelium. Cavity a follicle contains storage product of secretory epithelium called colloid. It is the only gland which stores its product outside the cells. When the gland is inactive there is a tendency for colloid to accumulate and for the epithelium to become low cuboidal to squamous or flattened. When the gland is overactive the colloid storage are depleted and the epithelium becomes columnar. The thyroid depleted and the epithelium becomes columnar. The thyroid follicles have a remarkable ability to trape inorganic iodine which can be stored and incorporated into hormones which are stored in the follicle cavity. But iodide concentrating activity of the thyroid gland is regulated by different control system.

TSH with preincubation of thyroid cell with iodide depressed, iodide concentration, On the other hand pretreatment with iodide depressed iodide pump activity as well as diminished the stimulatory effect of TSH added there after (Sherwin and Tong, 1974). According to Sinha and Singh (1990) extrathyroidal conversion of T4 to T3 increases during spawning phase as compared to prespawning phase in fresh water catfish *Clarias bastrachus* during its annual reproductive cycle. Melandar and Sundler (1972) suggested that in rat intrathyroid mast cells participate in the process of thyroid hormone secretion through effect of substrate released from these cells upon stimulation by TSH. The process of synthesis of thyroid hormones are same in fishes as in higher vertebrates.

The thyroid hormones are thyroxine (tetraiodothyronine) and triiodothyronine. The follicular cells synthesize a protein thyroglobulin which is present in colloid. Thyrogobulin is an important as subsequent synthesis of thyroid hormone. Valenta *et al.* (1982) suggested that intracellular thyroglobulin iodination stimulates rat thyroids. In response to endocrine stimulation, follicular cells engulf the colloid by phagocytosis. The colloid within the endocytotic vesicle is enzymatically degraded to yield thyroid hormones which are released from follicular cells into the extracellular space where they enter the abundant capillaries.

The first demonstration of thyroxine synthesis by the teleostean thyroid gland was in platyfish *Xiphophorus maculatus* (Berg and Gorbman, 1953) and the rate of turnover of 131 I–labelled thyroxine and triodothyronine in adult frogs and toads is temperature dependent (Dowling *et al.*, 1964). The thyroid function is regulated by thyrotrophic hormones, which are released by teleostean pituitary. The increase in quantity and activity of follicle is only due to continuous stimulation of thyroid gland by TSH. This result in colloidal content and cells became columnar in shape and increased in size. Where as in absense of TSH synthesis of thyroid hormone is minimum as a result the follicular cells become, flattened and lumen remained enlarged and full of colloid.

Many effect of thyroid hormone have been described in teleosts however its exact physiological role is still not fully known. Thyroxine had also been claimed to increase common excretion by goldfish (Hoar, 1958). Treatments with thyroxine leads to silvering of the skin in many teleosts (Pickford and Atz, 1957). According to various workers thyroxine stimulate the growth in many teleosts. Thyroxine also stimulates the formation of bone in various teleosts and sturgeon (Gorbman, 1969). It plays an important role in regulation of annual, gonadal and body weight cycle of birds (Hohn, 1961; Thapliyal, 1969, 1978, 1980, 1981; Assenmacher, 1973; Assenmacher and Hallageans. 1980; Assenmacher and Jallageas, 1980; Thapliyal *et al.*, 1982, 1983; Lal and Thapliyal, 1984). Lal and Thapliyal (1984) suggested that in migratory Redheaded Bunting *Emberiza bruniceps*, certain levels of thyroid hormones are required for full growth and coupling of annual gonadal and body weight cycles. Thyroid hormones also plays an important role in photorefractory state of the migratory Redheaded Bunting *Emberiza brunicep*.

From the above evidences it can be concluded that in fishes thyroid plays an important role during osmoregulation, accelerating metabolic rate, during formation of bone, silvering of skin, during full growth. However its exact role in a fish *Channa orientalis* is still not known. Certain drugs inhibit thyroid function by anatagonizing formation of thyroid hormones. But De Escober and Del Rey; (1962) stated that thiourea do not depress the peripheral deiodination of thyroid hormones in rat. According to Das and Isichei (1988) hypothyroidism cause structural changes in the thyroid cell membrane which would interfere with the trapping of iodine and shedding of thyroxine in the hypothyroid patients in human beings. The thyroid follicles have remarkable ability to trap inorganic iodine, which can be stored and incorporated into hormones, which stores into follicular cavity. But the antithyrid drug thiourea causes inhibition of iodine incorporation into tyrosine. But the action of thiourea on histology of thyroid gland are not still recorded. Therefore an attempt has been made to study the effect of antithyroid drug thiourea on the morphology and histology of the thyroid gland.

An attempt therefore has been made to clarify this point further by subjecting the fish *Channa orientalis* (Present study) to thyroidectomy. But here the surgical thyroidectomy is not possible because of the diffused structure of the gland. So chemical thyroidectomy had been tried with the help of antithyroid drug thiourea.

Materials and Methods

The species which has been selected for the present work is of economic value and readily available throughout the year in market, in the rivers and water bodies, nearby Amravati region. Locally the fish is known as "DOK" and is a common edible fish in the region. The fishes (*Channa orientalis*) were acclimatized to laboratory conditions for two weeks according to APHA/AWWA/WPCF (1975) standard methods.

In the present experiments the animals were treated with high and low concentration of thiourea solution.

Two hundred and twenty four young, matured ecclimatized animals were selected having similar sizes and weights. The animals were segregated into three groups.

Group 1 (Total 112 fishes)

As control group which is further subdivided into group IA and group IB having 56 animals in each group. Fishes received injections of fish saline (0.7 per cent).

Group II

As high concentration group for thiourea, in which fishes received intramuscular injections of 3 per cent thiourea (dose–0.3 mg/gm of body weight) at alternate day (total 56 fishes). Thiourea was prepared in 0.7 per cent Nacl solution.

Group III

As low concentration group for thiourea in which fishes received intramuscular injections of 2 per cent thiourea (dose–0.2 mg/gm of body weight) at alternate day (total 56 fishes).

Group IA used as control for group II and group IB is used as control for group III.

For studying histological changes in glands, fishes from groups–I, II, III were sacrified by decapitation at the interval of 24, 48, 72, 96, 168, 360 and 720 hours. At sacrifice lower jaws with thyroid were decalcified in 6 per cent solution of formic acid in 5 per cent formaldehyde for about a week. The tissues were dehydrated by passing through various grades of alcohol, cleared in xylene and a embedded in paraffin wax (56–58 c). Blocks were prepared and sectioned at 4– 6 μ and stained with following techniques.

Lower jaw with thyroid were stained with Ehrlich haematoxylin-eosin technique.

Assessment of Thyroid Activity

Assessment of thyroid activity was measured by microscopic examination of transverse sections (5–6 μ) of lower jaws in both control and experimental groups. This was supplemented by measurements in the following parameters.

1. Average number of follicles in a section (5 to 6 section/ fish).
2. Average follicular cell height was calculated by measuring 5 to 6 cells per follicles at random in each of the 4 to 5

follicles per section. Thus measurements were carried out in 5 sections per animal. This average epithelial cells height were calculated of animal's thyroid gland.

3. Average size of the follicles were estimated by measuring diameter of 5 to 6 follicles per section and 5 sections per animal. And thus the mean diameter of follicles was found out of the animal.
4. Average number of cells per follicles were counted random follicle per section. This procedure was followed for 5 follicles per section and 5–6 slides per animal tissue, yielded the average number of cells per follicle.

The data obtained after various measurements were tested for statistical significance. The experimental values were statistically evaluated by student's "t" formula.

Results and Observations

Thiourea Treated Fish Thyroid

Histology of Thyroid Exposed to High (0.3 mg/gm of body weight) Dose Exposure (Plate 25.1–A, B, C, D, E, F; Table 25.1; Figure 25.1)

Thyroid exposed to thiourea exhibited marked histological changes. The follicles were increased in number from 28.412 (Control) to 144.252 at 360–720 hours exposure (Table 25.1). Follicles became irregular in shape (Plate 25.1C) The diameter of follicles observed was 59.662 μ in control group, but decreased to 32.814 μ, 29.655 μ, 20.502 μ, 19.488 μ at the interval of 24–48 h, 72–96 h, 168–360 h, 360–720 h exposure respectively. The height of follicles increased from 3.970 μ in control group to 4.931 μ (24–48 h), 5.104 μ (72–96 h) 5.933 μ (168–360 h) and 6.155 μ (360–720 h). But the number of cells per follicles showed a significant reduction. Percent changed values are noted in Table 25.1 of the cells per follicles were 36.201, –44.822, –64.942, –68.133 after 24–48, 72–96, 168–360 and 360–720 hours of exposure respectively. In some follicles cells lining it multiplies and projected into the follicular cavity (Plate 25.1E) Nuclei became pycnotic. In some follicles loss of colloidal material was observed (Plate 25.1E). Vacuolation in the cytoplasm of secretory cells was also noticed (Plate 25.1F). The follicles were found closely packed and widely distributed in the lower jaw region.

Table 25.1: Effect of Thiourea Treatment on the Thyroid Gland (Histological Parameters) of the Fish *Channa orientalis* (High Dose –0.3 mg/gm of Body Weight)

Experimental Period (h)	*Follicles Per Section*	*Cell Height (μ)*	*Follicular Diameter*	*Cells Per Follicles*
Control	28.412 ± 2.706	3.970 ± 0.252	59.662 ± 3.166	34.814 ± 3.496
24–48	59.022*** ± 5.702	4.931* ± 0.301	32.814** ± 2.011	22.210** ± 1.922
	(107.742)	(24.182)	(–45.022)	(–36.201)
72–96	82.531 NS ± 5.530	5.104*** ± 0.270	29.655 NS ± 1.708	19.221*** ± 1.923
	(109.592)	(28.462)	(–50.302)	(–44.822)
168–360	119.302*** ± 5.622	5.933* ± 0.330	20.502* ± 1.877	12.210* ± 1.464
	(320.077)	(–49.377)	(–65.633)	(–64.942)
360–720	144.252 NS ± 5.774	6.155*** ± 0.387	19.488 NS ± 1.513	11.099 NS ± 1.416
	(407.920)	(54.911)	(–67.344)	(–68.133)

P values: *: < 0.1, **: < 0.01; ***: < 0.001.

NS: Non Significant, ( ): Parenthesis figures are percentage change.

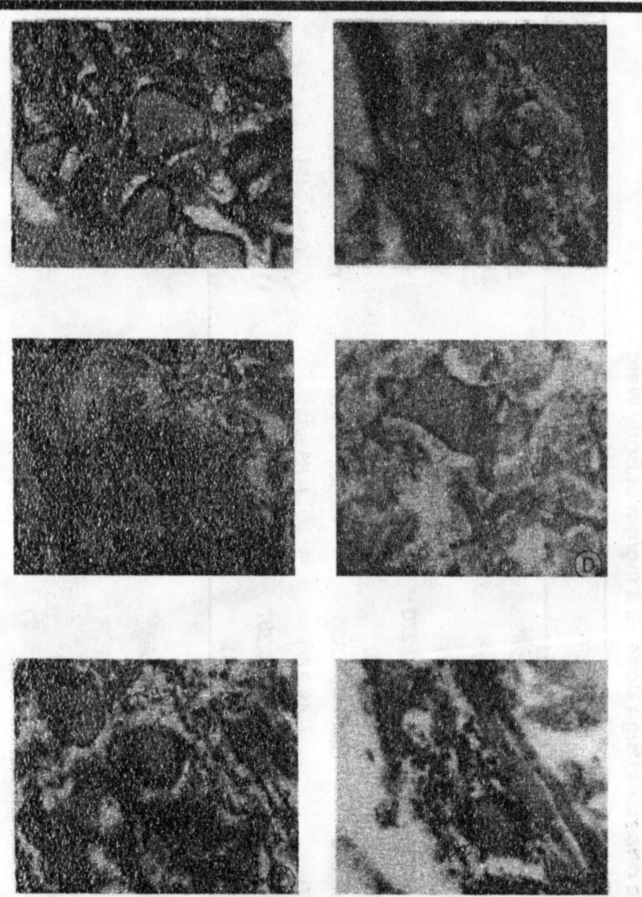

Plate 25.1: Histological Changes in Thyroid of *Channa orientalis* Exposed to Thiourea (High Dose)

A: T.S. of control *Channa orientalis* thyroid; B: T.S. of thiourea exposed to *C. orientalis* thyroid during early hours (24–48h); C: T.S. of thiourea exposed to *C. orientalis* thyroid during middle hours (72–96h); D: T.S. of thiourea exposed to *C. orientalis* thyroid during middle hours (168h); E: T.S. of thiourea exposed to *C. orientalis* thyroid during late hours (360h); F: T. S. of thiourea exposed to *C. orientalis* thyroid during late hours (720h).

NU: Nucleus; CO: Colloid; FE: Follicular epithelium; TF: Thyroid follicles; BG: Blood vessels.

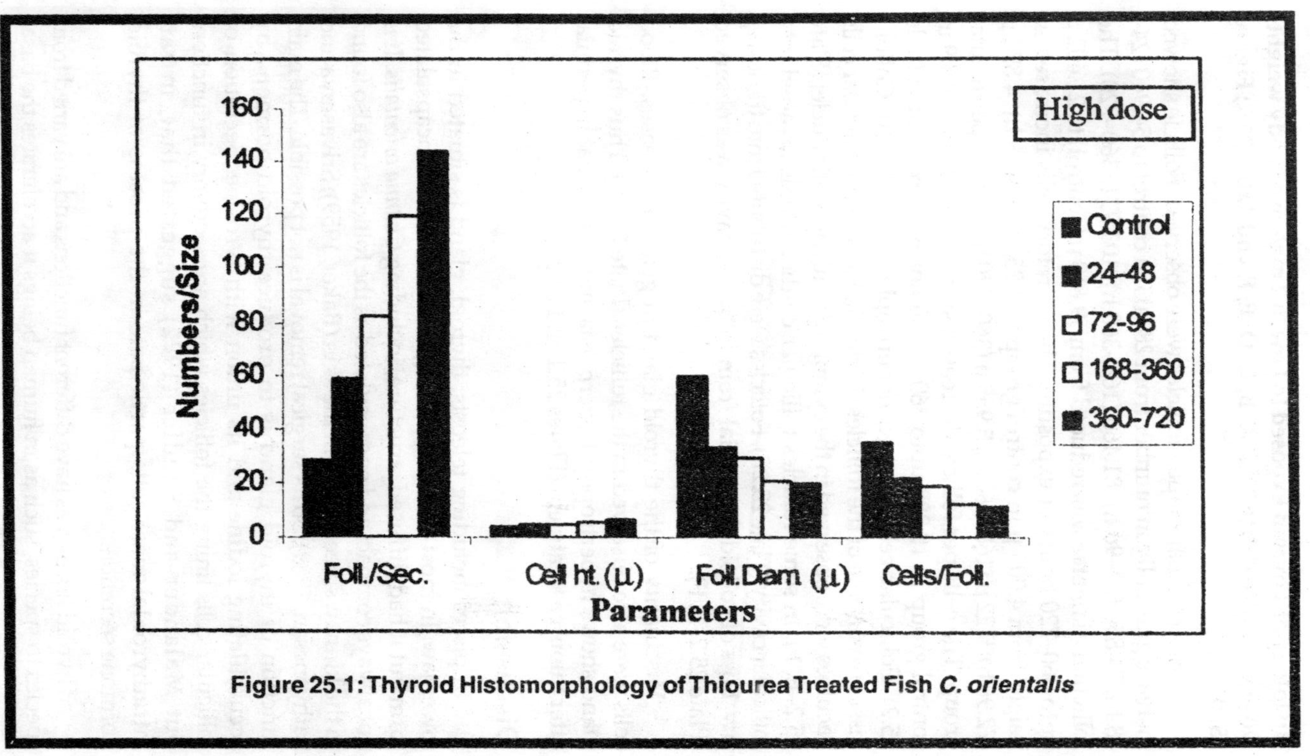

Figure 25.1: Thyroid Histomorphology of Thiourea Treated Fish *C. orientalis*

Histology of Thyroid Exposed to Low (0.2 mg/gm of body weight) Dose Exposure (Plate 25.2–A, B, C, D, E, F, and Table 25.2; Figure 25.2)

Some irregular shaped follicles were observed. Follicle showed increase gradually in number from 28.403 in control to 54.940 (24–48 h); 74.850 (72–96 h); 81.460 (168–360 h); 103.954 (360–720). The follicular diameter was reduced from 59.613 μ (control) to 21.402 μ upto 360–720 hours of exposure. The secretory cells increased in height from 4.003 μ in control group to 4.233 μ (24–48 h); 4.872 μ (72.96 h); 5.22 μ (168–360) h; 5.955 μ (360–720 h) in the experimental group. The cells per follicle decreased significantly from 34.809 in control group to 19.387 upto 360 to 720 hours of exposure (Table 25.2). Follicular cells showed hypertrophy (Plate 25.2B). Colloid losses from some of the follicles (Plate 25.2E) In some follicular cells vaculoes were observed in the cytoplasm with pycnotic nuclei (Plate 25.2–E, F). In some follicles follicular epithelial cells invaded into follicular cavity (Plate 25.2E). Follicles were distributed into the lower jaw. Loss of colloidal material from follicular cavity was observed (Plate 25.2–E, F).

Vascularity of the thyroid gland in general increases. Blood cells were also observed in the follicles (Plate 25.2E). Thus thyroid gland shows hypertrophy, hyperplasia and some what hypermia, after thiourea treatment (Plates 25.1 and 25.2).

Discussion

In pisces including teleosts, thyroid, gland is situated in the lower jaw. In most teleost thyroid gland is not in an encapsulated form but it had diffused structure, including *Channa orientalis*. It is not always compact and encapsulated as the follicles are also found in the kidney, spleen, even in the skin (Baker, 1959); Srivastava and Sathyanesan, 1969). So its surgical removal is not possible. The main function of thyroid gland is to produce thyroid hormone by accumulating iodine and its union with tyrosine. Secretion of follicular cells lining the follicles which are secretory in function. But Malander and Sundler (1972) suggested that, in rat, intrathyroidal mast cells participate in the process of thyroid hormone secretion.

Thyroid hormones have different functions and are varied from species to species, such as in human beings it accelerates the basal

Table 25.2: Effect of Thiourea Treatment on the Thyroid Gland (Histological Parameters) of the Fish *C. orientalis* (low dose –0.2 mg/gm of body weight)

Experimental Period (h)	*Follicles Per Section*	*Cell Height (μ)*	*Follicular Diameter*	*Cells Per Follicles*
Control	28.403 ± 2.724	4.003 ± 0.264	59.631 ± 3.152	34.809 ± 3.485
24–48	54.940 NS ± 4.142	4.233** ± 0.236	46.711 NS ± 2.244	25.812* ± 2.382
	(93.430)	(5.495)	(–21.666)	(–25.846)
72–96	74.850*** ± 4.632	4.872** ± 0.221	36.532 NS ± 2.233	24.402** ± 2.071
	(163.528)	(21.708)	(–38.652)	(–29.897)
168–360	87.460** ± 5.188	5.224** ± 0.284	26.701** ± 2.202	21.354** ± 2.144
	(207,925)	(–30.502)	(–55.207)	(–38.653)
360–720	103.954 NS ± 6.200	5.955 ± 0.286	21.402** ± 2.192	19.387*** ± 2.102
	(265.996)	(48.763)	(–64.109)	(–44.304)

P values: *: < 0.1, **: < 0.01; ***: < 0.001.

NS: Non Significant, ( ): Parenthesis figures are percentage change.

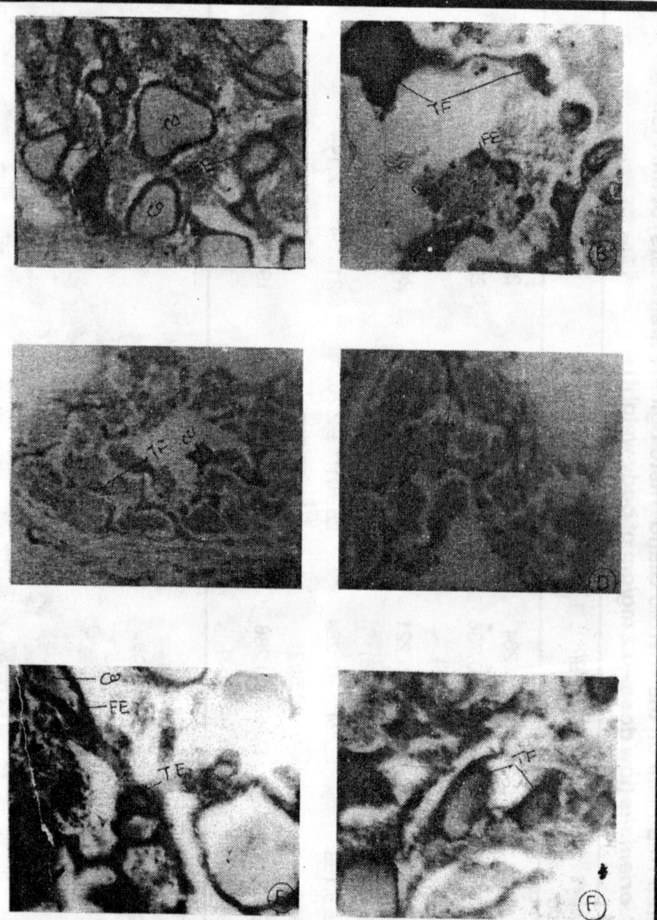

Plate 25.2: Histological Changes in Thyroid of *Channa orientalis* Exposed to Thiourea (Low Dose)

A: T.S. of control *Channa orientalis* thyroid; B: T.S. of thiourea exposed to *C. orientalis* thyroid during early hours (24–48h); C: T.S. of thiourea exposed to *C. orientalis* thyroid during middle hours (72–96h); D: T.S. of thiourea exposed to *C. orientalis* thyroid during middle hours (168h); E: T.S. of thiourea exposed to *C. orientalis* thyroid during late hours (360h); F: T. S. of thiourea exposed to *C. orientalis* thyroid during late hours (720h).

NU: Nucleus; CO: Colloid; FE: Follicular epithelium.

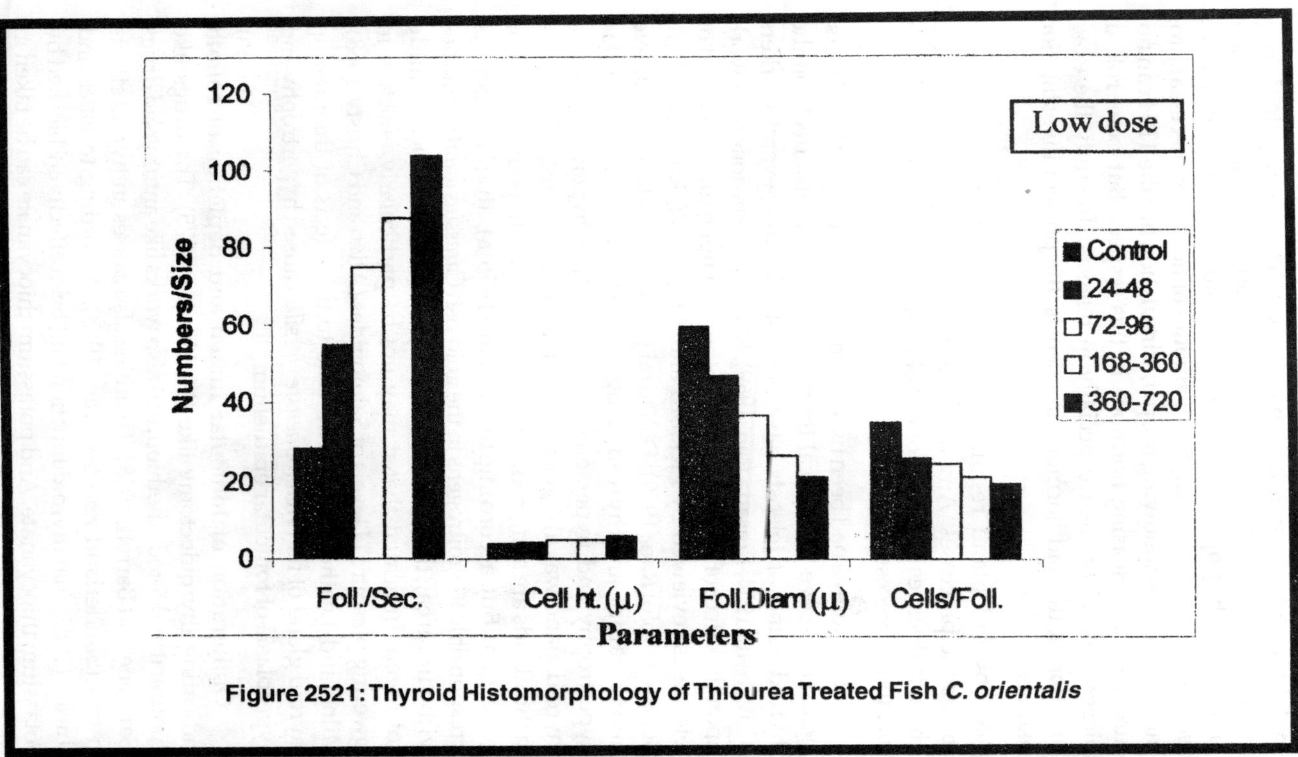

Figure 2521: Thyroid Histomorphology of Thiourea Treated Fish *C. orientalis*

metabolic rate, in amphibians, it plays important pole during reproduction, fall of growth in birds. According to Darrs and Kuhn (1982), the hypophysis and the thyroid gland of *Ambystome mexicanum* may release optimal amount of hormones necessary for metamorphosis following proper, stimulation but the TRH can not functions, as releasing hormone in this respect. But exact role in pieces is still not clear especially in teleosts. To study the exact function of thyroid hormone in teleosts, the present investigation had been made.

Thyroid gland regulates the organs, organ systems and metabolic processes. Any disturbance in the secretion of thyroid gland influences the functioning of organ, organ system as well as metabolic processes.

The thiourea had been used as an antithyroid drug (Pitt. Rivers, 1950) to achieve "Chemical thyroidectomy" and its effects on the thyroid gland of teleosts have been studied by several workers (Olivereau, 1954; Prasada Rao, 1969). Many workers done their work upon the effect of antithyroid drug thiourea upon thryroid gland of teleosts is reviewed by Pickford and Atz. (1967); Kinnear (1960); Belsare (1965); Rao (1969); Dehadrai (1971). According to all above investigators antithyroid drug *i.e.* thiourea caused cellular hypertrophy and hyperaemeia and follicular hyperplasia in the thyroid gland was also recorded in teleosts. All above alternations in the glands are symptoms or characteristics of hypothyroidism or goitrogen. But according to Chavin (1956 a), thyroid gland is irresponsible for goitrogens in the gold fish *Carassius auratus*. Studies with thiourea, the most commonly employed thiocarbamide goitrogen, indicated that it may not be particularly effective in lowering plasma TH levels of Salmonides (Allen and Christy, 1968); Milne and Leatherland, 1978). Due to the effects of thiourea on thyroid gland of fish *Heteropneustes fossilis* caused hypertrophy and hyperplasia of follicular epithelium.

Obliteration of follicular lumen and depletion of colloid indicating thyroidectomy like (Mukherjee, 1975). The drugs like thiouracil and 5-iodothothiouracil also works like antithyroid drugs (Sarclone and Barrett, 1958). Thiamine also causes anthyroid effects in rats (Slinderland and Sullivan; 1968). According to Eales and Shirley (1982) that thyroid function had been effectively blocked by potassium thiocyanate. And potassium thiocyanate can be taken in evaluating the effectiveness of thyroidal inhibitors in teleosts.

In the present investigation thyroid histological picture of *C. orientalis* showed increase in height of follicular cells and hyperaemia, (signs of hyperactivity) non-vacuoluted colloid, (hypo function), atrophic changes in the thyroid gland *i.e.* cyst formation and fibrosis. Thyroid follicles invaded in the extrathyroidal areas of lower jaw.

Thus the present thiourea treated fishes showed profound changes in the thyroid gland. Activeness or inactiveness of the gland reflected in the form of follicular cell size. The height of the follicular epithelium had been shown to be a criterion of activity of the cells in the thyroid gland of mammals, reptiles, amphibians and fishes (Uotila, 1939; Maity, *et al.*, 1973; Uhlenhuth, *et. al.*, 1945; Stolk, 1951). In Indian fruit bat *Rousettus* after antithyroid drug propylthiouracil treatment causes the increase in height of follicular cells. Lumen of the follicles contained less colloid and numerous vacuoles at the peripheral region of colloid which indicated hyperactive stage of thyroid (Shrikhande, 1983). Similar histological changes in the thyroid were observed following an antithyroid drug propylthiouracil administration (Gorbman and Bern, 1974; and Turner and Bagnara, 1976).

The active thyroid gland showed tall columnar epithelium, while in the inactive thyroid gland with reduced function showed squamous type of epithelium of follicular cells (Gorbman and Bern, 1962). The follicular epithelium of *Heteropneustes fossilis* is squamous but after treatment with thiourea the height of the epithelium increased (Mukherjee; 1975).

Other than the structure of follicular cells, the condition of colloid of follicles is also most important. Appearance of vacuoles in the colloid has been shown to be a sign of the thyroid activity and a high rate of absorption of the colloid from the follicles (Gorbman and Bern, 1962). According to Pickford and Atz (1957) the presence of numerous vacuoles in the colloid expresses high activity and their reduction or absence indicates proportional or total inactivation of the gland. Thiourea treated *Heteropneustes fossilis* showed numerous vacuoles in the colloid of the thyroid gland which due to quick absorption of thyroid hormone from the colloid of the follicles. Many follicles became acolloida, their nuclear size reduced and thus gland become atrophied in the fish *Heteropneustes fossilis* (Mukherjee,

1975). Atrophy of thyroid follicles after treatment with thiourea was also reported in *Phoxinus phoxinus* (Barrington and Matty, 1955). Thus the size and number of vacuoles in the colloid has been found increase in correlation with that of the follicular epithelial height (Mukherjee, 1975). Thiourea treated fish showed non-vacuolated colloid suggests reduced thyroid function.

References

Allen, D.M. and M. Christy (1978). Thiourea does not block visual pigment responses to prolactin in trout. *Vision Res.*, 18: 859–860.

APHA, AWWA and WPCF (1975). *Methods for Examination of Water and Wastewater*, 14th ed. Am. Publ. Hlth. Asso. Washington.

Assenmacher, I. (1973). In: *Avain Biology*, Vol. 3, (Eds.) D.S. Farner and J.K. King. Academic Press, New York, pp. 183.

Assenmacher, I. and M. Hallageas (1980). In: *Environmental Endocrinology*, (Eds.) Assenmacher and D.S. Farner. Springer Verlag, Berlin, pp. 52.

Assenmacher, I. and M. Jallageas (1980). Adaptive aspects of endocrine regulation in birds. In: *Hormones, Adaption and Evolution*, (Eds.) S. Ishii, T. Hirano, and M. Wada. Springer Verlag, Berlin/New York, p. 93–102.

*Baker, K.F. (1959). *Zoologica*, 44: 133.

Barrington, E.J.W. and A.J. Matty (1955). The identification of thyrotrophin secreting cells in the pituitary gland of the minnow (*Phoxinus phoxinus*). *Quart. J. Micr. Sci.*, 96: 193–201.

Belsare, D.K. (1965). Changes in thyroid gland, pituitary gland and gonads after thiourea treatment, in *Channa punctatus* Bloch. *Zool. Polon.*, 15: 231–245.

*Berg, O. and A. Gorbman (1953). *Proc. Soc. Exp. Biol. A. Med.*, 83: 751.

Chavin, W. (1956a). Thyroid distribution and function in the goldfish, *Carassius auratus* L., as determined by the uptake of tracer doses of radioiodine. *Anat. Rec.*, 124: 272.

Darras, V.M. and E.R. Kuhn (1982). Effects of TRH, Bovine TSH and Pituitary Extracts on thyroidal T4 Release in *Ambystoma mexicanum*. *Gen. Comp. Endocrinol.*, 51: 286–291.

Das, S.C. and U.P. Isichei (1988). Hypothyroidism and its Effects on thyroid issue and plasma lipids. *Indian J. Exp. Biol.*, 26: 92–94.

De Escobar, G.M. and F. Escobar Del Rey (1962). Influence of Thiourea, potassium perchlorate and thiocyanate and graded doses of prophylthiouracil on thyroid hormones metablism in thyroidectomized Rats, Isotoically, Equillibrated with varying Doses of Exogenous Hormone. *Endocrinology*, 71: 906–913.

Dehadrai, P.V. (1971). Thyroid in relation to aerial respiration in *Notopterus notopterus* Hamilton (Teleostei). *Biol. Sci.* (In Press).

Dowling, J.T., D. Razevska and D.J. Goodner (1964). Metabolism of thyroid hormone in Frogs and Toads. *Endocrinology*, 75: 157.

Eales, J.G. and S. Shirley (1983). Influence of potassium thiocyanate on thyroid function of Rainbow trout *Salmo qairdneri. Gen. Comp. Endocrinol.*, 51: 39–43.

Gorbman, A. (1969). Thyroid Function and its control in Fishes. In: *Fish Physiology*, Vol. II: The Endocrine System, (Eds.) W.S. Hoar and D.J. Randall. Academic Press, New York and London, p. 241–274.

Gorbman, A. and H.A. Bern (1974). *Textbook of Comparative Endocrinology*. Wiley Eastern Private Ltd., pp. 114.

Gorbaman, A. and H.A. Bern. (1962). *A Test Book of Comparative Endocrinology*. John Wiley and Sons Inc., New York, pp. 468.

Hoar, W.S. (1958). Effect of Synthetic thyroxine and gonadal steroids in the metabolism of goldfish. *Canad. J. Zool.*, 36: 113–121.

Hohn, E.O. (1961). Endocrine glands, thymous and pineal body. In: *Biology and Comparative Physiology of Birds*, Vol. 2, (Ed.) A.J. Marshall. New York, Academic Press, pp. 87.

Kinnear, J.E. (1960). Minimum dosage of thiourea effectively inhibiting. Thyroxine synthesis in the flounder. (*Platichthyes stellatus*). *Canad. J. Zool.*, 38: 917–922.

Lal, P. and J.P. Thapliyal (1984). Photorefractoryness in the migratory red headed bunting. *Proc. 1st Intern. Symp. Environ. and Horm*, (Eds.) Ischii, B.K. Follett and A. Chandola. Springer Verlag, Berlin (In press).

*Maity, L.K., A. Ghosh and S.K. Banerjee (1973). *Indian J. Exp. Biol.*, 11: 34.

Melandar, A. and F. Sundler (1972). Significance of thyroid mast cells in thyroid Hormone secretion. *Endocrinology*, 90: 802–807.

Milne, R.S. and J.F. Leatharland (1978). Effect of ovine TSH Thiourea, ovine prolactin and bovine growth hormone on plasma thyroxine and triiodothyronine levels in rainbow trout (*Salmo gairdneri*). *J. Comp. Physiol.*, 124: 105–110.

Mukheriee, A. (1975). Effects of thiourea treatment on thyroid and ovary of the catfish *Heteropneustes fossilis* (Bloch.). *Indian. J. Exp. Biol.*, 13: 327–332.

*Olivereau, M. (1954). A cited by Pickford and Atz., 1957.

Pickford, G.E. and J.W. Atz. (1957). *The Physiology of the Pituitary Gland of Fishes*. Zoological Society, New York, pp. 613.

*Pitt-Rivers, R. (1950). *Physiol. Rev.*, 30: 194.

*Rao, P.D.P. (1969). Effects of thiourea–treatment on the thyroid and pituitary gland of *Heteropneustes fossilis* (Bloch.). In: *Advance Notes on Symposia and Discussions, Proc. Indian Sci. Congr. 56th Session*, Bombay.

Sherwin, J.R. and W. Tong (1974). The actions of iodide and TSH on thyroid cells showing duel control system for the iodide pump. *Endocrinology*, 94: 1465.

Shrikhande, D.V. (1983). An experimental analysis of certain pituitary–adrenal gonadal relationship in Indian fruit bat *Rousettus lesshenaulti*. *Ph.D. Thesis*. Nagpur University, Nagpur.

Sarclone, E.J. and H.W. Barret (1958). The antithyroid activity of 5-iodo-2-thiouracil. *Endocrinology*, 63: 143–150.

Slingerland, D.W. and J.J. Sullivan (1968). An antithyroid Effect thiamine. *Endocrinology*, 82: 895–897.

Stolk, A. (1951). Histoendocrinological analysis of gestation phenomenon in the cyprinodont *Lebistes reticulatus* Peters. I. Thyroid activity during pregnancy. *K. Ned. Acad. Wet. Proc.* C.54: 550–557.

Thapliyal, J.P. (1969). Thyroid in avain reproduction. *Gen. Comp. Endocrinol, Suppl.*, 2: 111–122.

Thapliyal, J.P. (1978). Reproduction in Indian birds. *Pavo*, 16: 151–161.

Thapliyal, J.P. (1980). Thyroid in reptiles and birds. In: *Hormones, Adoption and Evolution* (Eds.) S. Ishii, T. Hirone and M. Wada. Japan Sci. Soc. Press Tokyo, Springer Varlag, Berlin, pp. 241–250.

Thapliyal, J.P. (1981). Presidential address, endocrinology of avian reproduction. In: *Proceedings, 68th Session of Indian Sci. Congr., Sec. Zoo. Entomol. and Fish.*, p. 1–11.

Thapliyal, J.P., A.K. Pati, V.K. Singh and P. Lal (1982). *Gen. Comp. Endocrinol.*, 46: 325.

*Thapliyal, J.P., P. Lal, A.K. Pati and B.B.Pd. Gupta (1983). *Gen. Comp. Endocrinol.*, 51: 444.

Turner, C.D. and Bagnara, J.T. (1976). *General Endocrinology,* 6th Edition. W.B. Sanders Company, Philadelphia.

*Uhlenhuth, E., J.E. Schenthal, J.U. Thompson and R.L. Zwilling (1945). *J. Morph.*, 76: 45.

*Uotila, V.V. (1939). *Endocrinology*, 25: 605.

Velenta, S.J., W.C. Florsheim and B.S. Sharma (1982). Acute effects of iodine on the stimulated rat thyroid. *Endocrinology*, 111: 1721–1727.

---

*: Not seen in original.

Index

D

I

J

K

L